Studies in Computational Intelligence

Volume 789

Series editor

Janusz Kacprzyk, Polish Academy of Sciences, Warsaw, Poland
e-mail: kacprzyk@ibspan.waw.pl

The series "Studies in Computational Intelligence" (SCI) publishes new developments and advances in the various areas of computational intelligence—quickly and with a high quality. The intent is to cover the theory, applications, and design methods of computational intelligence, as embedded in the fields of engineering, computer science, physics and life sciences, as well as the methodologies behind them. The series contains monographs, lecture notes and edited volumes in computational intelligence spanning the areas of neural networks, connectionist systems, genetic algorithms, evolutionary computation, artificial intelligence, cellular automata, self-organizing systems, soft computing, fuzzy systems, and hybrid intelligent systems. Of particular value to both the contributors and the readership are the short publication timeframe and the world-wide distribution, which enable both wide and rapid dissemination of research output.

More information about this series at http://www.springer.com/series/7092

Roger Lee
Editor

Software Engineering Research, Management and Applications

Editor
Roger Lee
Software Engineering and Information
Technology Institute
Central Michigan University
Mt. Pleasant, MI, USA

ISSN 1860-949X ISSN 1860-9503 (electronic)
Studies in Computational Intelligence
ISBN 978-3-030-07544-6 ISBN 978-3-319-98881-8 (eBook)
https://doi.org/10.1007/978-3-319-98881-8

This Springer imprint is published by the registered company Springer Nature Switzerland AG
The registered company address is: Gewerbestrasse 11, 6330 Cham, Switzerland

Foreword

The purpose of the 16th International Conference on Software Engineering, Artificial Intelligence Research, Management and Applications (SERA 2018) held on June 13–15, 2018 at Kunming, China is aimed at bringing together scientists, engineers, computer users, and students to share their experiences and exchange new ideas and research results about all aspects (theory, applications, and tools) of Software Engineering Research, Management and Applications, and to discuss the practical challenges encountered along the way and the solutions adopted to solve them. The conference organizers selected the best 17 papers from those papers accepted for presentation at the conference in order to publish them in this volume. The papers were chosen based on review scored submitted by members of the program committee and underwent further rigorous rounds of review.

In Chapter "Temporal Locality with a Long Interval: Hybrid Memory System for High-Performance and Low-Power", Bo-Sung Jung and Jung-Hoon Lee designed a DRAM and PCM hybrid memory system with low power consumption and high performance based on effective temporal locality. They proposed a page management method based on the temporal locality of a write reference center. According to the simulation results, the proposed hybrid memory achieved performance improvement from about 13 and 10% from energy-delay product compared with CLOCK-DWF and CLOCK-HM.

In Chapter "Design and Evaluation of a MMO Game Server", Youngsik Kim and Ki-Nam Kim implement a simple MMO Game Server using IOCP and evaluates its performance. Also, IOCP packet design and processing method are presented. The simple MMO Game Server implemented in this paper also supports multi-thread synchronization and dead reckoning.

In Chapter "Automatic Generation of Image Identifiers Based on Luminance and Parallel Processing", Je-Ho Park, Young B. Park, and Mi-Eun Ko propose a method to construct indexing of images utilizing the concept of the luminance area. The experimental evaluation of the proposed method illustrates that the proposed method satisfies the requirements for the image identification while reducing the processing cost.

In Chapter “Interface Module for Emulator-Based Web Application Execution Engine”, Hyunwoo Nam and Neungsoo Park propose a modified web-based emulator with the interface module and API, called as the web emulator-based execution engine. The experimental emulator-based web application was implemented and tested to evaluate the overall system.

In Chapter “A Study on the Influence and Marketing Effect of Korean Wave Events and Festivals Organization”, Jae Ho Park, Jeong Bae Park, and Cheong Ghil Kim introduce a method of measuring the influence and marketing effect of organizations for Korean Wave Events and Festivals. The feasibility of measuring results is also ensured by exploratory factor analysis.

In Chapter “Understanding the Success Factors of R&D Organization”, Donghyuk Jo and Jongwoo Park attempt to understand the contributing factors of Research & Development (R&D) project in terms of social capital perspective, which is being considered key resource of business management today. The significance of this study is in validating the importance of team capital and competence building under R&D project environment and presents strategic direction for R&D project success and team competence.

In Chapter “Study on Detection Algorithm of Live Animal in Self-bag-Drop Kiosk in Airport Using UWB Radar”, Kiwon Jung, Younghwan Bang, and Sun-Myung Hwang proposed a method of detection by UWB (Ultra-Wide Band) to prevent against safety accidents which could be occurred in Self-Bag Drop installed and unmanned operated in airport by the unexpected intrusions such as alive animals, humans especially.

In Chapter “A Study on Success Factors for Business Model Innovation in the 4th Industrial Revolution”, Sung-Hwan Yoon, Nguyen Si Thin, Vo Thi Thanh Thao, Eun-Tak Im, and Gwang-Yong Gim conduct an analysis that will be made to see what factors are important for unicorn enterprises. While the existing studies emphasized regulations and entrepreneurship aspects, the unicorn enterprises that are currently governing the world have been realized by having innovative business models as the key competence with entrepreneurship and regulations added.

In Chapter “A Study on the Efficiency of Global Major Mobile Operators”, Jeongil Choi1, Youngju Park, and Yonghee Kim analyzed the efficiency of global major mobile operators and the reason for their efficiency. For this purpose, the financial data of 96 operators in 40 OECD member countries were utilized. Based on this financial data, this study conducted a comparative analysis of the efficiency among operators and among countries.

In Chapter “A Study on Effects of Supporting Born Global Startups Policy Affecting the Business”, Jung-Ran An, SungTaek Lee, Ju-Hyung Kim, and Gwang-Yong Gim focused on corporate finance and nonfinancial performance. This study not only provides meaningful information on the implementation of policy research and the implementation of startup policy, but it also provides a framework for the study.

In Chapter "Design of the Model for Indoor Location Prediction Using IMU of Smartphone Based on Beacon", Jae-Gwang Lee, Seoung-Hyeon Lee, and Jae-Kwang Lee propose a room location prediction model that can improve user's position accuracy and detect user's position in case of signal loss using Beacon and smartphone sensor.

In Chapter "IoT Implementation of SGCA Stream Cipher Algorithm on 8-bits AVR Microcontroller", Mouza Ahmed Bani Shemaili, Chan Yeob Yeun, Mohamed Jamal Zemerly, Khalid Mubarak, Hyun Ku Yeun, Yousef Al Hammadi, and Yoon Seok Chang designed a lightweight and secure stream ciphers for IoT to secure hardware and software that can fit constrain resources devices. Thus, they implement their proposed solution on 8-bit AVR microcontroller in order to study the required memory and speed.

In Chapter "A Study on Upgrading Non-urban Areas-Using Big Data the Case of Hwang Ze and Danggok Districts", Yong Pil Geum examines areas with potential future growth in resident populations through political implication from population expansion in non-urban areas. Using big data, Hwang Ze and Danggok districts in Gyeongsangbuk-do, South Korea, were chosen as research subjects owing to their proximity to Jillyang-eup of Gyeongsain-si, where rural and industrial areas are in contact and urbanization is taking place for upgrading non-urban area using big data.

In Chapter "Simulation of Flood Water Level Early Warning System Using Combination Forecasting Model", Kristine Bernadette Barrameda, Sang Hoon Lee, and Su-Yeon Kim explore the use of BPNN and SVM techniques as a combined model using the Minimum Variance (MV) method to predict the upcoming flood water level events in Calinog River, Iloilo, Philippines.

In Chapter "A Study on the Components that Make the Sound of Acceleration in the Virtual Engine of a Car", Sang-Hwi Jee, Won-Hee Lee, Hyungwoo Park, and Myung-Jin Bae study the virtual engine sounds that can enhance the feeling of acceleration by controlling the playback speed of a virtual engine sound. An MOS test showed that the virtual engine sound was not much different from the engine sound of an existing engine.

In Chapter "A Study on the Characteristics of an EEG Based on a Singing Bowl's Sound Frequency", Ik-Soo Ahn, Bong-Young Kim, Kwang-Bock You, and Myung-Jin Bae analyzed the sound of a singing bowl, which is used as a method to restore and maintain the balance of the natural frequency of the human body, and studied the EEG (electroencephalogram) of the listener according to the frequency band of the singing bowl's sounds.

In Chapter "A Study on the Stability of Ultra-High Frequency Vocalization of Soprano Singers", Uk-Jin Song, Ik-Soo Ahn, Myung-Sook Kim, and Myung-Jin present a study on the Stability of Ultra-High Frequency Vocalization of Soprano Singers. To confirm whether these soprano singers actually show distinct vibrations in the high frequency range, they analyze the vibration characteristics of four Korean sopranos to ascertain the depth of their vocal vibration.

It is our sincere hope that this volume provides stimulation and inspiration, and that it will be used as a foundation for works to come.

Program Chairs

Nanjing, China

Bing Luo
Nanjing University

Chengdu, China

Junfeng Wang
Sichuan University

Kunming, China
June 2018

Zhengtao Yu
Kunming University of Science and Technology

Contents

Contributors

Ik-Soo Ahn Sori Engineering lab, Department of Information and Telecommunication Engineering, Soongsil University, Dongjak, Seoul, South Korea

Yousef Al Hammadi College of Information Technology, UAE University, Al Ain, UAE

Jung-Ran An Department of Business Administration, Soongsil University, Seoul, Republic of Korea

Myung-Jin Bae Sori Engineering Lab, Department of Information and Telecommunication Engineering, Soongsil University, Sangdo-Dong, DongJak-Gu, Seoul, South Korea

Younghwan Bang Korea Institute of Industrial Technology, Chungcheongnam-do, Republic of Korea

Mouza Ahmed Bani Shemaili CIS Division, HCT, Ras al Khaimah, UAE

Kristine Bernadette Barrameda School of Computer and Information Engineering, Daegu University, Gyeongsan, Republic of Korea

Yoon Seok Chang School of Air Transport and Logistics, Korea Aerospace University, Goyang, South Korea

Jeongil Choi College of Business Administration, Soongsil University, Seoul, South Korea

Yong Pil Geum Catholic University of Daegu, Gyeongsan, South Korea

Gwang-Yong Gim Department of Business Administration, Soongsil University, Seoul, Republic of Korea

Sun-Myung Hwang Daejeon University, Daejeon, Republic of Korea

Eun-Tak Im Soongsil University, Seoul, Korea

Sang-Hwi Jee Department of Telecommunication Engineering, Soongsil University Sori Engineering Lab Sangdo-Dong, Dongjak-Gu, Seoul, Korea

Donghyuk Jo Department of Business Administration, Soongsil University, Seoul, South Korea

Bo-Sung Jung Department of Control and Instrumentation, Gyeongsang National University, Jinju, Gyeongnam, Korea

Kiwon Jung SCom CNS Inc, Daejeon, Republic of Korea

Bong-Young Kim Department of Information and Telecommunication, Soongsil University, Dongjak, Seoul, South Korea

Cheong Ghil Kim Department of Computer Science, Namseoul University, Cheonan, Choongnam, Korea

Ju-Hyung Kim Department of IT Policy and Management, Soongsil University, Soongsil University, Seoul, Republic of Korea

Ki-Nam Kim Department of Game and Multimedia Engineering, Korea Polytechnic University, Siheung-si, Republic of Korea

Myung-Sook Kim Department English Language and Literature, SoongSil University, Sangdo-Dong, DongJak-Gu, Seoul, Korea

Su-Yeon Kim School of Computer and Information Engineering, Daegu University, Gyeongsan, Republic of Korea

Yonghee Kim College of Business Administration, Soongsil University, Seoul, South Korea

Youngsik Kim Department of Game and Multimedia Engineering, Korea Polytechnic University, Siheung-si, Republic of Korea

Mi-Eun Ko School of Computer Engineering, Hansung University, Seoul, South Korea

Jae-Gwang Lee Department of Computer Engineering Hannam University, Daejeon, Korea

Jae-Kwang Lee Department of Computer Engineering Hannam University, Daejeon, Korea

Jung-Hoon Lee ERI, Department of Control and Instrumentation, Gyeongsang National University, Jinju, Gyeongnam, Korea

Sang Hoon Lee School of Computer and Information Engineering, Daegu University, Gyeongsan, Republic of Korea

Seoung-Hyeon Lee Information Security Research Division, ETRI, Daejeon, Korea

Sung Taek Lee Department of IT Policy and Management, Soongsil University, Soongsil University, Seoul, Republic of Korea

Won-Hee Lee Department of Telecommunication Engineering, Soongsil University Sori Engineering Lab Sangdo-Dong, Dongjak-Gu, Seoul, Korea

Khalid Mubarak Dubai Men's College, HCT, Dubai, UAE

Hyunwoo Nam Department of Computer Science and Engineering, Konkuk University, Seoul, Korea

Hyungwoo Park Department of Telecommunication Engineering, Soongsil University Sori Engineering Lab Sangdo-Dong, Dongjak-Gu, Seoul, Korea

Jae Ho Park Department of Performance Planning and Management, ChungWoon University, Hongseong, Choongnam, Korea

Je-Ho Park Department of Software Science, Dankook University, Yongin, South Korea

Jeong Bae Park Department of Performance Planning and Management, ChungWoon University, Hongseong, Choongnam, Korea

Jongwoo Park Department of Business Administration, Soongsil University, Seoul, South Korea

Neungsoo Park Department of Computer Science and Engineering, Konkuk University, Seoul, Korea

Young B. Park Department of Software Science, Dankook University, Yongin, South Korea

Youngju Park Graduate School of Business, Soongsil University, Seoul, South Korea

Uk-Jin Song Sori Engineering Lab, Soongsil University, DongJak-Gu, Seoul, South Korea

Vo Thi Thanh Thao Soongsil University, Seoul, Korea

Nguyen Si Thin Soongsil University, Seoul, Korea

Chan Yeob Yeun ECE Department, Khalifa University of Science and Technology, Abu Dhabi, UAE

Hyun Ku Yeun NS Division, HCT, Abu Dhabi, UAE

Sung-Hwan Yoon Soongsil University, Seoul, Korea

Kwang-Bock You Department of Information and Telecommunication, Soongsil University, Dongjak, Seoul, South Korea

Mohamed Jamal Zemerly ECE Department, Khalifa University of Science and Technology, Abu Dhabi, UAE

Temporal Locality with a Long Interval: Hybrid Memory System for High-Performance and Low-Power

Bo-Sung Jung and Jung-Hoon Lee

Abstract In this paper, the main idea is to design DRAM and PCM hybrid memory system with low power consumption and high performance based on effective temporal locality. PCM has two major drawbacks by write operation. First, the number of write operations is limited. Second, PCM has a longer write operation time than DRAM. On the other hand, DRAM can effectively overcome the disadvantages of PCM. Therefore, a page replacement algorithm suitable for the characteristics of DRAM and PCM is necessary for effective DRAM and PCM hybrid memory operation. For page management considering the characteristics of PCM, we proposed a page management method based on the temporal locality of a write reference center. In this paper, pages with the temporal locality that are referenced at short intervals will be managed with hybrid memory; pages with temporal locality that are referenced at long intervals are managed by the proposed buffer system. Furthermore, a hot page is defined by a write operation and a buffer system, and this page is managed in DRAM to reduce the overhead of PCM. According to the simulation results, the proposed hybrid memory achieved performance improvement from about 13 and 10% from Energy-delay product compared with CLOCK-DWF and CLOCK-HM.

Keywords Hybrid memory · Memory architecture · Temporal locality
Low-power · Performance

B.-S. Jung
Department of Control and Instrumentation, Gyeongsang National University,
501 Jinju-Daero, Jinju, Gyeongnam, Korea
e-mail: blueking80@gnu.ac.kr

J.-H. Lee (✉)
ERI, Department of Control and Instrumentation, Gyeongsang National University,
501 Jinju-Daero, Jinju, Gyeongnam, Korea
e-mail: leejh@gnu.ac.kr

R. Lee (ed.), *Software Engineering Research, Management and Applications*, Studies in Computational Intelligence 789, https://doi.org/10.1007/978-3-319-98881-8_1

1 Introduction

With the advent of the diverse social networking services and the evolution of IOT systems, the latest applications with high capacities and data concentration are emerging [1, 2]. As a result, the working-set of the current computing system is increasing more and more. It is also indispensable to increase the capacity of the memory for data processing and storage.

Today DRAM is widely used as a main memory of a computing system as an advantage such as low cost, high speed memory access time and access to a byte address. However, DRAM can no longer function the role of the ideal of memory due to the limitation of degree of integration and high consumption power [3, 4]. The main memory, DRAM, accounts for roughly 40% of the total energy consumption of the entire system [5]. Most of this power consumption is used for refresh operation of capacitor and the transistor leakage current.

Currently, new memory structures and algorithms using next generation nonvolatile memory have been studied to overcome the limitations of DRAM. PCM, STT-Ram and RRam are attracting attention as next-generation nonvolatile memories [6, 7]. In particular, PCM is guaranteed reliability and better performance than RRam, and it has attracted attention as a substitute memory of DRAM due to higher integration density than STT-RAM. Especially PCM has 4 times higher integration density than DRAM. It also has byte-address access like DRAM. However, there are following problems in using PCM as main memory [8–10]. First, the number of writes is limited to 10^7–10^8 like NAND flash. Secondly, for write operations, it has high energy consumption and delay time (5–10 compared to DRAM).

One of the best ways to overcome the problems of PCM is the hybrid memory structure that operates with DRAM. In this hybrid structure, the write-intensive page is managed in the DRAM, thereby minimizing the problem of the write operation of the PCM [11]. Therefore, this paper proposes a high performance-low power DRAM and PCM hybrid memory system. The key idea in this paper is, first, that the proposed hybrid system is a separation of hot-page and cold-page, taking into account efficient memory operation and PCM write limitations. One of the ways to improve performance of an effective memory system is to use two localities in program execution. In this paper, we extended the concept of temporal locality. In other words, the page that accessed the hybrid memory is assumed to be likely to be referenced again in the near future. Second, frequent DRAM and PCM page swapping has greater performance degradation than PCM write operations. Therefore, in this paper, page replacement between DRAM and PCM occurs only in pages selected from hot-page from PCM to replace pages of DRAM and PCM.

Selective data buffering [12], Hot-cold data Filtering [8], and DRAM write-cache [13] proposed a DRAM buffer structure with various block sizes for PCM. This research reduced PCM access by storing sub-blocks that are more likely to be referenced in DRAM buffers of various block sizes. However, these operations are sequentially accessed from the DRAM buffer to the PCM.

CLOCK-DWF [10] and M-CLOCK [14] has improved the performance of DRAM and PCM hybrid memory systems by managing the pages where frequently request a write reference in DRAM. However, in order to operate DRAM and PCM memories simply for write requests, this research is unable to predict characteristics at the time of program execution. CLOCK-HM [15] and Migration-Based [16] proposed a method of operating a DRAM and PCM hybrid memory by predicting characteristics under execution of a program. However, in order to judge the page exchange, the page has a disadvantage that it is necessary to continuously update the boundary value each time a reference occurs.

In this paper, we compare the performance of hybrid memory systems (CLOCK-DWF, CLOCK-HW) with the same memory configuration and purpose. Simulation results show that the energy-delay product, which is a performance index considering the proposed hybrid average memory access time and energy consumption, has an average improvement of 10 and 13%, respectively, compared with CLOCK-DWF and CLOCK-HW.

2 Proposed Hybrid Memory System

2.1 Motivation

In order to reduce write reference of PCM in hybrid memory, it is effective to store as many pages as possible to refer to writing to dram. Prediction of pages those are likely to be referred to when executing this program is important. Using DRAM for unilateral writing reference pages can rather cause various problems. A Write and a read reference cannot be accurately predicted at program execution. Furthermore, depending on the characteristics of the program, the ratio of write reference and read reference is not constant. Therefore, the most important consideration for effective operation of hybrid memory is dram's effective memory operation.

Utilizing two localities for characteristics at the time of execution of a program is as one of effective methods for improving the performance of the memory system. The spatial locality means that the possibility that adjacent data of recently referenced data is likely to be referenced, temporal locality means that recently referenced data is likely to be referenced again in the near future. In general, spatial locality is sensitive to block size, whereas temporal locality is sensitively to the number of blocks. The basic unit of the current main memory system is 4 KB in the page, page management considering temporal locality rather than spatial locality can be effective.

Another effective way to predict the characteristics of pages during program execution are to use the information on the page when a page fault is occurring. This makes it easy to manage pages with temporal locality that has a long time gap. Therefore, if the page extracted from the recent memory is frequently requested, it is effective to manage the page from memory for a long time. However, it is difficult to use such information directly from lowest hierarchical memory. Therefore, in order to utilize the page extracted from the recent memory, an additional history buffer is necessary.

As a result, managing pages based on the temporal locality of the memory system is an effective way of improving the performance. However, temporal locality with only reference to the conventional page is difficult to solve the problem of PCM in DRAM and PCM hybrid memory. Therefore, we considered the temporal locality of the write reference for DRAM and PCM hybrid memory system.

In this paper, we newly defined and predicted temporal locality of a write reference for effective page management. In other words, pages where write references occur frequently are intensively managed from the DRAM. Furthermore, after these pages are extracted from the hybrid memory, if the page is requested in the near future, the page is determined to be temporal locality and managed from the DRAM. We call the hot page.

2.2 The Proposed Algorithm

In this paper, the proposed Hybrid memory system is a memory of the same layer in both DRAM and PCM and is basically executed based on the CLOCK algorithm [13]. And the proposed Hot_page implementation can be roughly divided into three situations.

(1) when the requested page refers to writing by a miss.
(2) when a page extracted from the recent hybrid memory is requested.
(3) Frequent write reference to a page requested.

The Hot_pages that defined by reference to requested writes and frequent page references can be determined during program execution. On the other hand, the Hot_page defined from the page extracted from the recent hybrid memory requires an additional list. Therefore, in this paper, we propose a victim list to exploit the information of recently extracted pages from hybrid memory system. This victim list will have information on hot pages defined from PCM along with the Meta information of pages discarded from hybrid memory system The Fig. 1a shows the state of the current hybrid memory system.

(1) **DRAM Operation**: Hot_page is stored in DRAM of the hybrid memory system, and basically DRAM will manage pages centered on write reference. For effective page management, DRAM has additional state bits. The proposed state bits are as follows.

Reference bit (R): This bit indicates a page in which a recent write reference or read reference has occurred, and is updated to '1' when a page reference occurs.
Write_bit (W): This bit indicates a page in which a recent write reference has occurred, and is updated to '1' when a write reference.
Pre_Write bit (PW): This bit indicates the previous state that the write bit is updated by the CLOCK-handle. That is, When the CLOCK-handle scans a page for victim

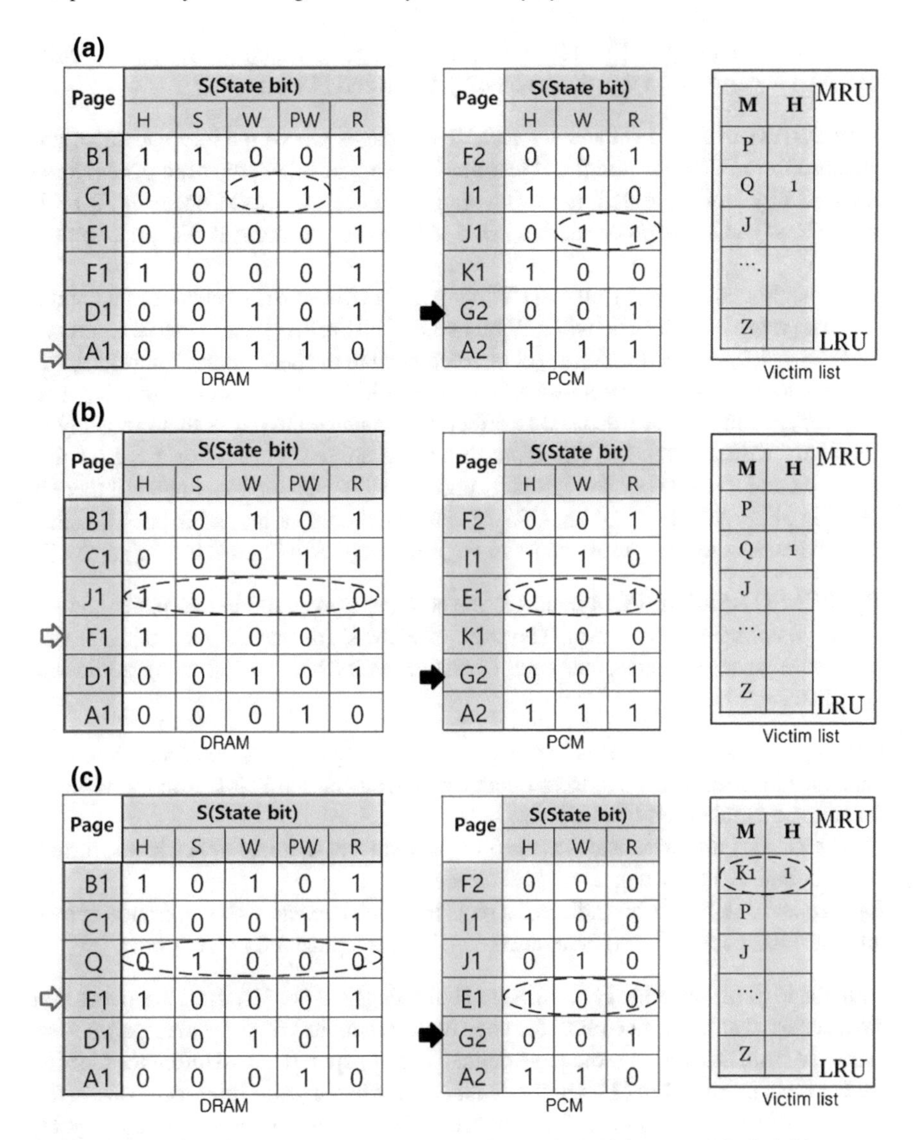
(a)

DRAM

Page	S(State bit)				
	H	S	W	PW	R
B1	1	1	0	0	1
C1	0	0	1	1	1
E1	0	0	0	0	1
F1	1	0	0	0	1
D1	0	0	1	0	1
A1	0	0	1	1	0

PCM

Page	S(State bit)		
	H	W	R
F2	0	0	1
I1	1	1	0
J1	0	1	1
K1	1	0	0
G2	0	0	1
A2	1	1	1

Victim list (MRU to LRU)

M	H
P	
Q	1
J	
….	
Z	

(b)

DRAM

Page	S(State bit)				
	H	S	W	PW	R
B1	1	0	1	0	1
C1	0	0	0	1	1
J1	1	0	0	0	0
F1	1	0	0	0	1
D1	0	0	1	0	1
A1	0	0	0	1	0

PCM

Page	S(State bit)		
	H	W	R
F2	0	0	1
I1	1	1	0
E1	0	0	1
K1	1	0	0
G2	0	0	1
A2	1	1	1

Victim list (MRU to LRU)

M	H
P	
Q	1
J	
….	
Z	

(c)

DRAM

Page	S(State bit)				
	H	S	W	PW	R
B1	1	0	1	0	1
C1	0	0	0	1	1
Q	0	1	0	0	0
F1	1	0	0	0	1
D1	0	0	1	0	1
A1	0	0	0	1	0

PCM

Page	S(State bit)		
	H	W	R
F2	0	0	0
I1	1	0	0
J1	0	1	0
E1	1	0	0
G2	0	0	1
A2	1	1	0

Victim list (MRU to LRU)

M	H
K1	1
P	
J	
….	
Z	

Fig. 1 The proposed hybrid memory system operation: **a** the initial states of the hybrid memory system, **b** page replacement operation, and **c** the page fault operation

page selection, if the W bit is ‘1’, the PW bit is updated to the W bit value and the W bit is updated to ‘0’.

***Hot_page* (*P*)**: This refers to a page where page replacement occurred in the DRAM due to frequent reference from the PCM. So, when pages are exchanged between DRAM and PCM, the H bit of the page stored in DRAM is updated to ‘1’.

***Second chance bit* (*S*)**: This bit represents the page in which the frequently write reference occurred, and it is determined by W bit and PW bit.

In DRAM of Fig. 1a, Page B1 and F1 are pages where page replacement has occurred form PCM because the H bit have '1'. And the other pages are pages stored in DRAM by Hot_page definition. If a read reference on DRAM occurred, R bit is updated to '1'. On the other hand, when a write reference occur R bit and W bit are set to '1'.

As mentioned earlier, W bit and PW bit indicate current write reference and previous write reference state. So when W bit and PW bit are all '1' and rewrite reference occurs, we defined that the page will be referenced in the near future. Therefore, it is effective to manage these pages for a long time with DRAM. Therefore, in this paper, S bit for second chance was used to effectively manage the pages that are likely to frequently write references. Therefore, if a write reference occurs on Page C1 that has W bit and PW bit are all '1', S bit is updated to '1'. On the other hand, if there is no '1' in either W bit or PW bit, DRAM is updated only W bit and R bit. The other hand, when a write reference occurs in Page E1, only W bit and R bit are set to '1'.

(2) **PCM Operation**: PCM manages the cold page proposed in this paper and the page extracted from dram. The page of PCM is managed by a read reference and a write reference. For such operation, PCM has the following additional bits.

***Hot_page bit* (*H*)**: This refers to a page in which page replacement occurred more than once in the DRAM due to frequent reference from the PCM. And the value of the state bit is inherited from DRAM.
***Write bit* (*W*)**: This bit indicates a page in which a recent write reference has occurred, and is updated to '1' when a write reference occurs.
***Reference bit* (*R*)**: This bit indicates a page in which a recent write reference or read reference has occurred, and is updated to '1' when a page reference occurs.

In the Fig. 1a, since the H bit of Page I1 and Page A2 are '1', these are pages that have occurred at least once more for page replacement to DRAM. Other pages were extracted from DARM, or Cold page defined in this paper. If a read reference Occurs, the R bit is simply set to '1'. On the other hand, when a write reference occurs, the state bits of W bit and R bit will be checked. If a write reference occurs on Page I1, both W bit and R bit are updated to '1'. On the other hand, if a write reference occurs on Page J1 with that W bit and R bit is '1', Page J1 is defined the Hot_page on PCM due to W bit and R bit. So, according to the proposed algorithm, PCM's Page J1 is a group of candidates for page replacement with dram's cold page.

(3) **Page replacement between DRAM and PCM**: If there is a free page for the page replacement, the page is saved in the free page. And H bit of the page that was stored in DRAM from PCM is updated to '1'. Then, the remaining state bits of DRAM are updated to '0'. And when page is saved from DARM to PCM, H bit and R bit of DRAM are inherited to H bit and R bit of PCM.

On the other hand, if the DRAM does not have a free page for page replacement, the victim page is selected by proposed algorithm. In this paper, DRAM is managed on the basis of the write reference. So, the victim page will select pages where the write reference related state bits are all '0'. That is, a page whose W bit and S bit which is state bits area all '0' is selected as a victim page. In order to select the victim page, in the Fig. 1a, we check the state bits of the W bit and S bit on Page A, which is indicated by the CLOCK_handle (white arrow). Since the W bit on Page A is '1', it is excluded from the candidates for the victim page. At this time, the PW bit is updated to '1' by W bit, and the W bit is updated to '0'. The CLOCK_handle indicate to page B1 which is the next page, and checks the state bits of the page. Page B1 is excluded from the candidates for victim page because S bit is '1', at this time, W bit is updated to '1' by S bit, and S bit is updated to '0'. Through this operation, Page C1 is also excluded from the candidate of the victim page. And Page E1 where W bit and S bit are all '0' is selected as the victim page. Then the CLOCK_handle indicate to Page F1 for the next state bit check. Here, selection of the victim page is executed only 1cycle for the page replacement between DRAM and PCM. If DRAM's victim pages are not selected during 1 cycle, there will be no page replacement between DRAM and PCM. Figure 1b show the state bits that is updated by to select victim page on DRAM, and the result of page replacement by DARM and PCM. In the PCM of the Fig. 1b, since H bit and R bit of the Page E1 that is selected a victim page are '1', H bit and R bit are inherited when page is exchanged to PCM. And H bit of Page J1 stored in the DRAM is updated to'1', and all the other state bits are updated to '0'.

(4) **Page fault operation**: When a reference failure occurs in a page in the Hybrid memory system, Hot-pages are stored in the DRAM according to the algorithm in which the requested pages are proposed from the memory of the lowest hierarchy, otherwise it is stored in PCM. Hot_page is defined as a page requested form write reference or extracted from recent hybrid memory system. Recent pages extracted from Hybrid memory means a pages existing in the victim list in this paper. This victim list is included an information as Meta data and hot page of page which is extracted from PCM.

If the Page Q is requested by a miss operation of Hybrid memory, the Page Q is stored in the DRAM because it is existing in the victim list. At this time, the victim page selection by the page fault is different form the page replacement between DRAM and PCM, the state bits of each pages are inspected until the victim page selected.

Because H bit of Page Q which is will be stored in DRAM from the victim list is '1', S bit is updated to '1' when Page Q is stored in DRAM. The H bit is the page where page replacement occurred between DRAM and PCM. Therefore, managing for a long time with DRAM in one of the ways that hybrid memory can reduce the page replacement. In Fig. 1a, since Page E1 is the victim page, Page Q is replaced with Page E1 and S bit is updated to '1'. In order to save the victim page E 1, the PCM will select the victim page by checking the W bit and R bit of each page. If R bit and W bit are all '0', the page will be selected as a victim page. Otherwise, the bits of each state are updated.

In PCM of the Fig. 1a, we will check the state bits from Page G2 where the CLOCK_handle(black arrow) is located. Page G2 is excluded the candidate group of the victim page because R bit of the pages is '1', and R bit is updated to '0' at that time. Also, Page A2 and Page F2 are excluded the candidate group of the victim page because R bit of the pages is '1', and R bit is updated to '0'. Even if the R bit of Page I1 is '0', it is excluded from the victim page candidate group because W bit is '1'. Here, when W bit is '1', the page means a candidate group of the page replacement between DRAM and PCM, which means a page with a high possibility of occurrence of rewrite reference. Therefore, it is effective that page for which a writ reference is requested exists for a long time in PCM. These operations are performed in the PCM that the victim page will be selected until.

In such an operation, Page K1 is selected as a victim page in PCM, where it is replaced by Page E1. And the CLOCK_handle for PCM point to Page G2 which is the next page of the replaced page. Then, the page K1 extracted from PCM is stored in the memory of the lowest hierarchy, and the meta data and H bit that is an information of the Page K1 are stored in the victim list. The Fig. 1c shows the result of page replacement when Page Q is defined in the Hot_page and the hybrid memories miss occurs.

If the requested Page Q is not in the victim list, the Page Q is replaced with the Page K1 of the victim page of PCM. Then all the state bits of the stored Page Q are updated to '0'. Here, the operation of the victim list is the same as the above-described operation.

3 Performance Evaluation

In this section, we evaluated the performance of the proposed hybrid memory system. For performance evaluation, we confirmed the memory access address of SPEC CPU 2006 through a modified Cachegrind tool from the Valgrind 3.6.3 toolset. We filter memory references out that are accessed directly from the cache memories and collect only 100 million main memory references. Therefore, the proposed hybrid memory system was performed by trace driven simulation.

And we also utilized the DRAM and PCM characteristics for the simulation, as shown in Table 1 [10, 17–19]. The default sizes of DRAM and PCM for hybrid memory system are 32 and 128 MB, respectively. Since the existing DRAM module is 128 MB, the size of PCM was also selected with the same size of 128 MB. And since the degree of integration of DRAM is 1/4 of PCM, it was chosen to be 32 MB.

3.1 *Victim List Size*

In this paper, Hot_page is defined by page which is requested for a write reference, and page which is existed in the victim list. The page which is requested write

Table 1 DRAM and PCM characteristics

Attribute	DRAM	PCM
Access granularity (byte)	64	64
Non-volatile	X	O
Read latency (ns)	50	50–100
Write latency (ns)	50	350
Read energy (nJ/p)	0.1	0.2
Write energy (nJ/p)	0.1	1.0
Endurance	X	10^8 for write
Idle power	~1.3 W/GB	~0.05 W
Density	Low	High (4XDRAM)

reference by a miss is defined by the characteristic of the program when the program is executed. On the other hand, Hot_page with pages extracted from recent hybrid memory are not. Therefore, the proposed Hot_page concept is more sensitive to victim list than write reference. In order to choose the size of an effective victim list, we measured the performance of hybrid memory system with various victim list sizes.

Because the write limit on PCM is also major problem of hybrid memory system, in this paper, we measured the number of a write time of PCM. Figure 2 shows the number of write operation to PCM of various sizes of victim list. As shown in Fig. 2, the beset performance is shown when the victim list is 25% of PCM size. This is because the count of pages moving from DRAM to PCM increases as the size of the victim list increases.

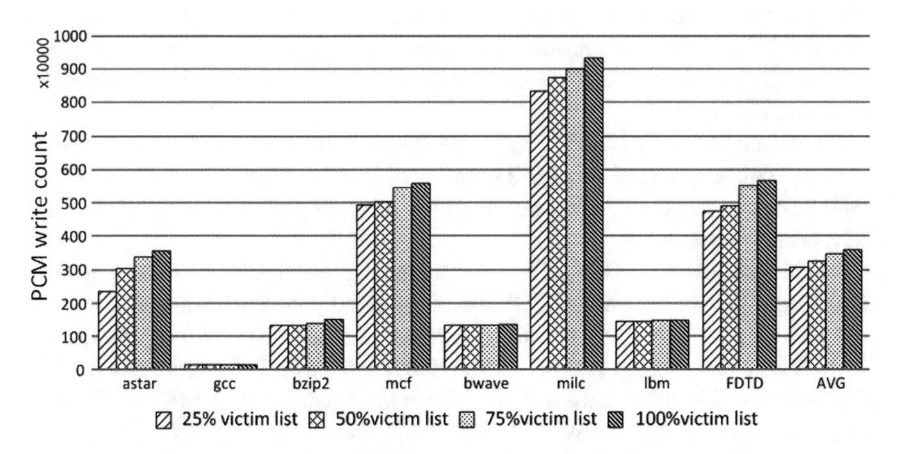

Fig. 2 PCM write count of the proposed hybrid memory with victim list of various sizes

3.2 Performance Evaluation

In order to evaluate the performance of the hybrid memory system in this paper, we compare and evaluate the performance with the existing researched hybrid memory system. We selected the CLCO-DWF and CLOCK-HW for comparison hybrid memory system. The CLOCK-DWF reduces the write reference of PCM by managing the write reference page of DRAM like the proposed hybrid memory system. However, the CLOCK-DWF only saves the write reference page in DRAM and the read reference page in PCM when the program is executed. This has the disadvantage that it cannot operate effective DRAM in a program in which a read reference page request frequently occurs. Also, when a write reference occurs to PCM, the page replacement occurs immediately in DRAM, so there is a problem of frequent a page replacement between DRAM and PCM.

As shown in this paper, CLOCK-HW is a structure using history buffer for managing pages extracted from hybrid memory. However the page in which the write reference frequently occurs is stored in DRAM, where as in the other pages, pages are randomly assigned to DRAM and PCM. Even though CLOCK-HW can efficiently use DRAM and PCM, it has the problem that it can effectively reflect the characteristics of DRAM and PCM.

In this paper, CLOCK-DWF and CLOCK-HW have the same size as the proposed hybrid memory system for performance evaluation. The Fig. 3 shows the write count of PCM. In the Fig. 3, the count of a write reference pages is basically 4 KB. In the proposed hybrid memory system and CLOCK-DWF, the lazy policy is applied when the page replacement between DRAM and PCM. Therefore, we measured the count of a write reference by lazy policy from them. Based on this, we converted it into a unit for a page write operation and added a number. That is, when the block size of the cache memory is 64 bytes, when the PCM block writes operation occurs 64 times, it is defined by one-page write operation. Simulation results show that the proposed hybrid memory system reduced the count of a write operation by about 10% compared to CLOCK-DWF, but showed a higher count of a write operation by 6% than CLOCK-HW.

The reason that CLOCK-DWF shows the lowest performance is that the page with which is requested a read reference by miss of hybrid memory is stored to PCM, and if a write request to the page occurs, the page replacement between DRAM and PCM occurs immediately.

On the other hand, CLOCK-HW can store the requested page in memory that can store it regardless of DRAM and PCM. In addition, the page can be moved to DRAM with PCM. Page migration does not occur from DRAM to PCM. For this reason, CLOCK-HW shows good performance improvement with PCM write counts. The proposed hybrid memory can effectively utilize memory compared to CLOCK-DWF in the intermediate stage between CLOCK-DWF and CLOCK-HW.

Figure 4 shows the average memory access time of hybrid memory systems. Unlike Fig. 3, CLOCK-HM shows the worst performance of average memory access time. As described above, CLOCK-HM does not cause page replacement between

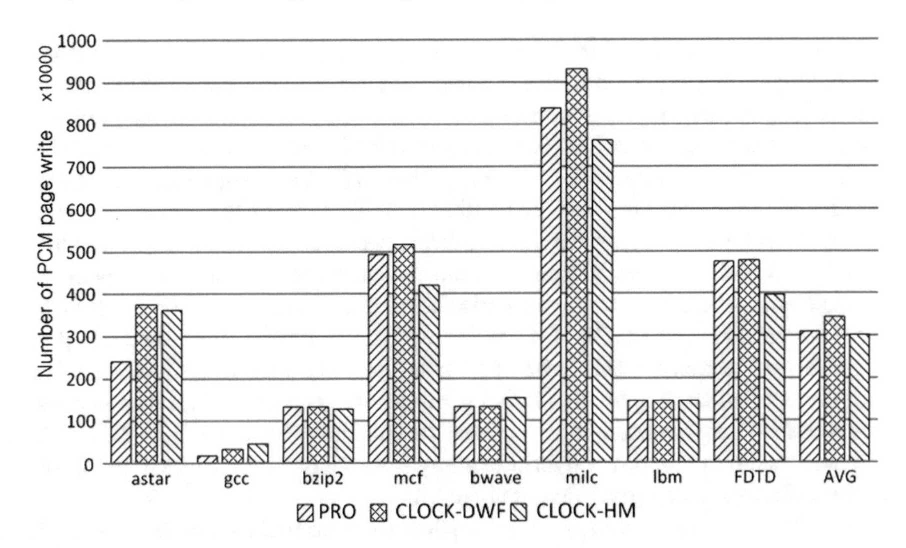

Fig. 3 PCM write count of the hybrid memory systems

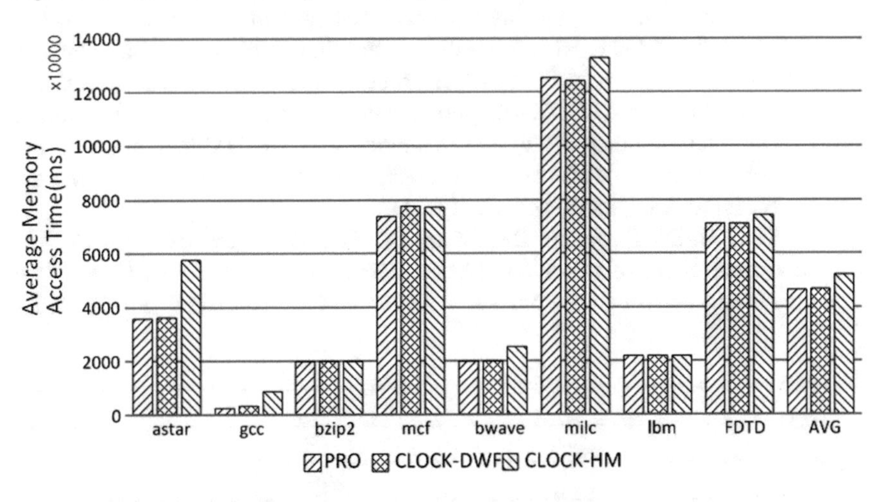

Fig. 4 Average memory access time of the hybrid memory systems

DRAM and PCM. Only the page where the write reference frequently occurs in the PCM is moved to the DRAM. Therefore, CLOCK-HM has a difficulty to manage page effectively than the proposed hybrid memory and CLOCK-DWF. In contrast, the proposed hybrid memory and CLOCK-DWF show similar average memory access times. The two hybrid memory structures have similar hit ratio. Also, when a write reference to the PCM occurs, pages are exchanged between DRAM and PCM.

According to the simulation results, the proposed hybrid memory improved average memory access time of at most 12%, compared to the CLOCK-HM.

Figure 5 is a diagram showing the energy consumption of the hybrid memory. According to the simulation results, CLOCK-HM shows the most effective performance improvement of energy consumption as a whole. CLOCK-HM simply moves the page where frequent writing has occurred to the DRAM. On the other hand, the proposed hybrid memory system and CLOCK-DWF have a page exchange operation between DRAM and PCM. According to Table 1, PCM write operation consumes about 10 times more energy than DRAM operation. However, page swapping between DRAM and PCM consumes 64 times more energy than the write operation by page unit. According to the simulation results, an average of 45% of the PCM write counters in the CLOCK-HM are 64 bytes of block write. Meanwhile, in proposed hybrid memory, block write of 64 bytes occurred only in 20% of the PCM write counter. According to the simulation results, the proposed hybrid memory system reduced energy consumption by about 10% compared with CLOCK-DWF. On the other hand, the proposed hybrid memory system compared with CLOCK-HM showed high energy consumption of about 3%.

In order to accurate performance evaluation, we measured energy-delay product, which is a performance index considering memory access time and energy consumption. Figure 6 shows the Energy-delay product of the hybrid memory system, which is normalized by the CLOCK-HM. According to the simulation results, the proposed hybrid memory has achieved the best performance more than CLOCK-DWF and CLOCK-HM. However, CLOCK-HM shows better performance with 'mcf', 'milc', and 'FDTD' than the other two hybrid memories. CLOCK-HM has improved the most effective performance of energy consumption despite having the lowest Average memory access time in "mcf", "milc", "FDTD".

Since CLOCK-DWF operates DRAM and PCM based on page operation, DRAM does not operate effectively even if DRAM size increases. Therefore, CLOCK-DWF shows the lowest performance with energy-delay product. Compared with CLOCK-

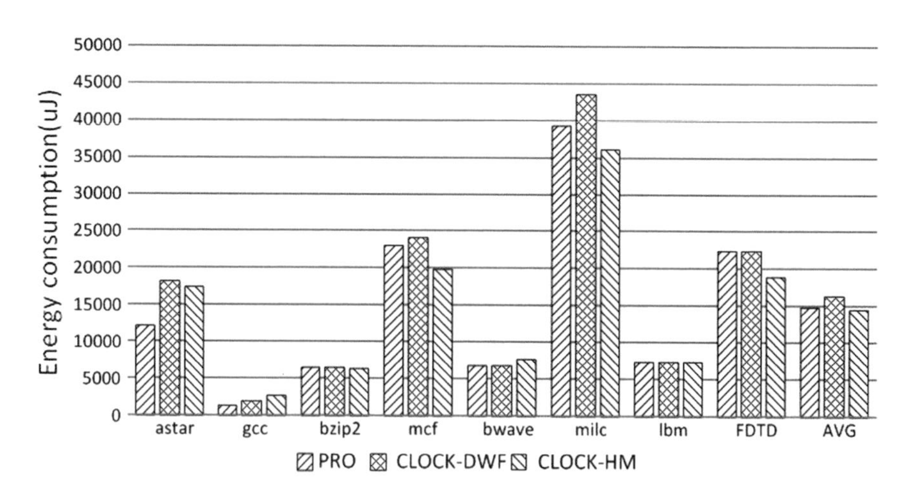

Fig. 5 Energy consumption of the hybrid memories

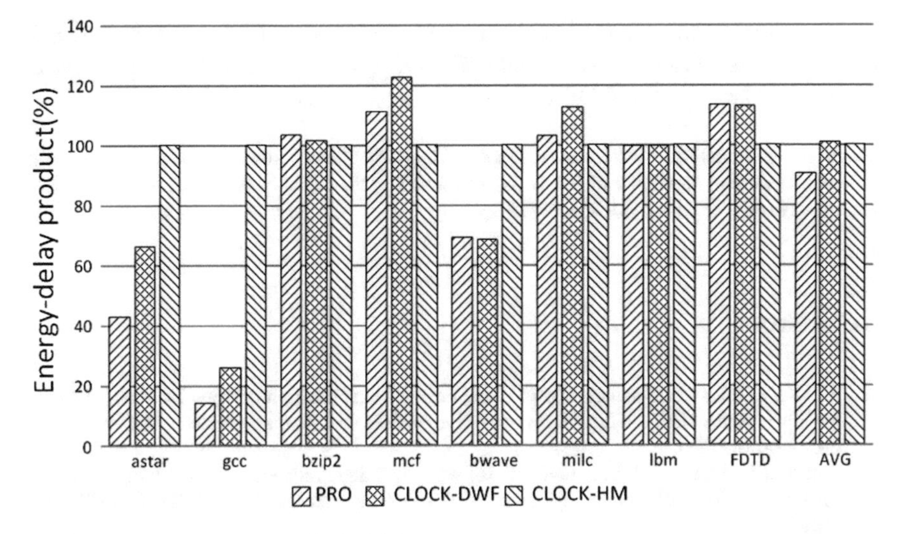

Fig. 6 Energy-delay product of the hybrid memory

DWF, the proposed hybrid memory can operate DRAM more effectively, and pages can be managed more than effectively CLOCK-HM. Therefore, we were able to realize the best performance improvement by energy-delay product which takes into consideration both the average memory access time and energy consumption.

According to the simulation results, the proposed hybrid memory achieved performance improvement from about 13 and 10% from Energy-delay product compared with CLOCK- DWF and CLOCK-HM.

4 Conclusion

PCM has attracted attention as a new alternative memory technology for reasons such as low power consumption and high integration density etc. This paper was proposed a new memory management technique for PCM and DRAM hybrid system with a low-power and a high performance.

This paper was newly defined a Hot_Page and a Cold_Page for hybrid memory system. The Hot page means the pages that requested for write reference or extracted from the recent hybrid memory. Otherwise it means the Cold-Page. To do this, in this paper, we are proposed the victim list for a Hot_page selection. And, considering the characteristics of DRAM and PCM, DRAM manages a Hot_page and PCM manages a Cold_page. In addition, pages are effectively managed by providing a second opportunity for pages that frequently send references.

According to the simulation results, the proposed hybrid memory structure achieved performance improvements of 10 and 13%, respectively, compared to CLOCK-HM and CLOCK-DWF.

Acknowledgements This research was supported by Basic Science Research Program through the National Research Foundation of Korea (NRF) funded by the ministry of Education, Science and Technology (NRF-2014R1A1A4A01008504).

References

1. Ma, K., Li, X., et al.: Incidental computing on IoT nonvolatiel processors. In: Proceedings of the 50th Annual IEEE/ACM International Symposium on Microarchitecture, MICRO-50 ‘17, pp. 204–218 (2017)
2. Sparsh, M., Jeffrey, S.V.: A survey of software techniques for using non-volatile memories for storage and main memory systems. IEEE Trans. Parallel Distrib. Syst. **27**(5), 1537–1550 (2016)
3. Prabhu, M., Rajarajan S., Suresh, K.S.: Proposed hybrid memory using DRAM and PCM to attain better performance. Am. Eurasian J. Sci. Res., 99–103 (2013)
4. Jang, S.I., Yoon, S.K., et al.: Data classification management with its interfacing structure for hybrid SLC/MLC PRAM main memory. Comput. J. **58**(11), 2852–2863 (2015)
5. Qureshi, M.K., Vijayalakshmi, S., Rivers, J.A.: Scalable high performance main memory system using phase-change memory technology. In: Proceedings of the 36th Annual International Symposium on Computer Architecture, pp. 24–33 (2009)
6. Palangappa, P.M., Li, J., Mohanram, K.: WOM-Code solutions for low latency and high endurance in phase change memory. IEEE Trans. Comput. **65**(4), 1025–1040 (2016)
7. Awad, A., Kettreing B., Solihin, Y.: Non-volatile memory host controller interface performance analysis in high-performance I/O systems. In: 2015 IEEE International Symposium on ISPASS, pp. 145–153 (2015)
8. Yoon, S.K., Jung K.S., et al.: Hot-cold data filtering and management for PRAM based memory-storage unified system. In: 2017 IEEE International Conference on Systems, Man, and Cybernetics (SMC), pp. 1609–1614 (2017)
9. Khouzani, H.A., Yang, C., Hu, J.: Improving performance and lifetime of DRAM-PCM hybrid main memory through a proactive page allocation strategy. In: 2015 20th Asia and South Pacific Design Automation Conference (ASP-DAC), pp. 508–513 (2015)
10. Lee, S.Y., Bahn, H.K., Noh, S.H.: CLOCK-DWF: a write-history-aware page replacement algorithm for hybrid PCM and DRAM memory architectures. IEEE Trans. Comput. **63**(9), 2187–2200 (2014)
11. Khouzani, H.A., Hosseini, F.S., Yang, C.: Segment and conflict aware page allocation and migration in DRAM-PCM hybrid main memory. IEEE Trans. Comput. Aided Des. Integr. Circuits Syst. **36**(9), 1458–1470 (2016)
12. Park, K.Y., Yoon, S.K., Kim, S.D.: Selective data buffering module for unified hybrid storage system. In: 14th International Conference on Computer and Information Science, pp. 173–178 (2015)
13. Chen, C., An, J.: DRAM write-only-cache for improving lifetime of phase change memory. In: International Midwest Symposium on Circuits and Systems (MWSCAS), pp. 1–4 (2016)
14. Lee, M.H., Kang, D.H., et al.: M-CLOCK: migration-optimized page replacement algorithm for hybrid DRAM and PCM memory architecture. In: Proceedings of the 30th Annual ACM Symposium on Applied Computing, pp. 2001–2006 (2015)
15. Cai, X., Ju, L., et al.: A novel page caching policy for PCM and DRAM of hybrid memory architecture. In: 13th International Conference on Embedded Software and Systems (ICESS), pp. 67–73 (2016)

16. Seok, H.C., Park, Y.W., Park, K.H.: Migration based page caching algorithm for a hybrid main memory of DRAM and PRAM. In: Proceedings of the 2011 ACM Symposium on Applied Computing, SAC '11, pp. 595–599 (2011)
17. Corbato, F.J.: A paging experiment with the multics system, in Honor of P.M. Morse, pp. 217–228. MIT Press, Cambridge (1968)
18. http://www.hpl.hp.com/techreports/2008/HPL-2008-20.html
19. Jiang, L., Zhang, Y., Yang, J.: Mitigating write disturbance in super-dense phase change memories. Proceedings of the 2014 44th Annual IEEE/IFIP International Conference on Dependable Systems and Networks, DSN '14, pp. 216–227 (2014)

Design and Evaluation of a MMO Game Server

Youngsik Kim and Ki-Nam Kim

Abstract Many large-scale online genre games such as Massive Multi-player Online Role Playing Game (MMORPG) are attracting attention in the game market. In a game server connected to hundreds or thousands of users, a large number of packets come and go between the server and the client in real time. For the server to endure these loads, IOCP (Input/Output Completion Port) and multi-thread are necessary. This paper implements a simple MMO Game Server using IOCP and evaluates its performance. Also, IOCP packet design and processing method are presented. The Simple MMO Game Server implemented in this paper also supports multi-thread synchronization and dead reckoning.

Keywords Massive Multi-player Online Role Playing Game (MMORPG) Game server · Multi-thread · IOCP (Input/Output Completion Port)

1 Introduction

MMOFPS (Massive Multi-player Online First Person Shooting) game in Fig. 1, (which has 100 users living on a large island in Korea recently), the Battle Ground, which was released at Blue Hall, had the shortest record of early access to the steam platform with sales of one million units in 16 days. Recently, the development of online games that are getting bigger than in the past is active. In the past, the development of large-scale games in more than 100 people has been a trend these days, compared to the 8-, 16-person online games. In a large online game, one server is responsible for as few as a hundred and as many as a thousand clients. The more clients connected to the server, the more network loads are placed on the server. In

Y. Kim (✉) · K.-N. Kim
Department of Game and Multimedia Engineering, Korea Polytechnic University, Siheung-si, Republic of Korea
e-mail: kys@kpu.ac.kr

K.-N. Kim
e-mail: ddous@kpu.ac.kr

R. Lee (ed.), *Software Engineering Research, Management and Applications*, Studies in Computational Intelligence 789, https://doi.org/10.1007/978-3-319-98881-8_2

Fig. 1 Screen shot of PLAYERUNKNOWN'S BATTLEGROUND2 related work

recent years, as almost everyone is using the Internet, there are also users who play online games. MMORPG(Massive Multi-player Online Role Playing Game) is the trend of the game which is recently released on mobile, and large-scale online games are also being released on PC and console side.

To design a large-scale online game server, IOCP (Input/Output Completion Port) [1] is almost essential in a Windows environment. This means that you will specify the port that will handle the completion of input and output. Since the notification at the completion time of input and output is processed by the overlapped input/output, this technique can be regarded as an extension of the superposition input/output technique of the window. A port is an object created to handle a task or service. It will be easy to recall that the port of a socket is an object for transferring data I/O to a particular service. Once you understand the characteristics of these ports, you can think about how to process the data by creating a port dedicated to notification at the completion of input and output. When I/O is completed from an input/output device (here, it is limited to a socket), this completion report is accumulated in the input/output completion queue. The thread wakes up, and the thread reads the completion report in the queue to process the data.

The advantage of this approach is that you can create a thread in advance and wait for I/O completion report, as opposed to the multithreaded way of creating a thread every time the client connects. By creating such a reasonable number of worker threads, the operating system can wake up idle worker threads. This approach is similar to a thread or process pool, but there is a difference. In general, it is not easy to allocate tasks to idle threads in the thread pool approach. To do this, I need to use some tricky techniques, and the IOCP does not need to worry about the developer, but the operating system picks up and wakes up the thread.

Multithreaded programming is inevitable to use IOCP. Multi-threading is not essential, but multi-threading is essential to building a game server using IOCP. Much work has been done to use multithreaded environment because synchronization is essential and it is the most important factor to improve performance. In this paper, we propose a basic server architecture and design to implement MMORPG and provide the following three techniques. First, we design and design the packet to be used in the MMORPG server. Second, we propose a synchronization technique for prevention of deadlock and data race prevention, and finally a dead reckoning technique for reducing the server load [2].

The game server is divided into a listener server and a dedicated server. The listener server runs as a process with the game client and can connect to a server hosted by another player without a separate server and play or invite other players. However, the disadvantage is that the server is also turned off when the game is turned off, so when the player ends the game, the host of the server is changed to another player. It is also a favorite way in LAN parties.

The dedicated server runs independently of the client process. It runs on a dedicated computer on a separate high-performance network, and players can enjoy pleasant gameplay by connecting to a dedicated server.

A dedicated server for large multi-user online games is a large-capacity game server operated by a specific company. You can run and maintain the server only by the developer who developed the game or by the developer's permission (so-called game publisher). In such a server, an unspecified number of players can enjoy the game together.

2 Related Works

2.1 *Structure of Game Server*

The first thing you have to decide before you develop any program is the programming language. In this paper, we propose C++ as a language for using IOCP. You can also build with C# or use .Net Frame Work with C#. .Net Frame Work is also implemented internally as IOCP, but the Garbage Collector in C# language has a big limitation. It is a big problem that there is a break in the game server where many packets are moving in real time. Using .Net Frame Work can increase productivity from the developer's point of view, but when it comes to commercialization, garbage-collector can cause server performance degradation. When using C++, it is possible to make low-level memory management compared to C#, so there is little overhead in writing and erasing the memory.

Multi-threaded use is inevitable to handle large amounts of packets received in real time from clients. The multi-threaded IOCP game structure [3] is generally in Fig. 2. Accept thread receives connections from clients through the thread. The connected client socket enters the asynchronous reception state after the information initialization operation registers it in the IOCP handle, and performs a full asynchronous I/O operation. When a packet is received for the socket, data is inserted

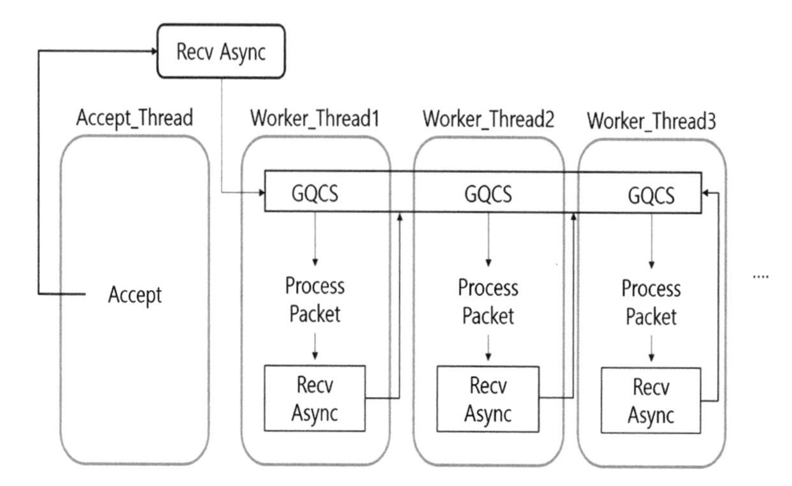

Fig. 2 Multi-threaded server architecture using IOCP

into the IO completion queue. Worker threads that are waiting are awakened by the GQCS (GetQueudCompletionStatus) function. If there is data in the queue, the GQCS function will wake up one of the worker threads randomly through the CPU scheduler, take out the working distance, and process the packet for the packet. The packet is processed and generated packets to be broadcast, and then the socket enters the asynchronous receive state. Figure 2 is a minimum structure for communication, and it is possible to add a thread to perform the function as needed by the developer.

In the case of a commercial game server, Fig. 3 has the form of a distributed server structure [4]. If you do not use the distributed server architecture, the load on the server increases, and as the number of users connected increases, you may see a common "lag" phenomenon, such as connection failure or delayed input. The client tries to connect through the login server, and the login server compares the client's information through the DB server and then starts communication with the main server. In the main server, user information is stored in the database in real time through DB Server. In the MMO Game, you have to operate hundreds of thousands of NPCs, and when you are in charge of the main server, the load on the server will grow exponentially. Therefore, separation of AI Server from MMORPG is essential. Figure 3 shows a functionally distributed server [5, 6]. Such a server has an advantage that the inconvenience caused by the service failure is reduced because the functions other than the functions handled by the suspended server are operating normally.

2.2 *Packet Design and Processing*

TCP conveys a byte stream that does not have a record boundary concept. TCP does not have the concept of "packets" that users can see. This can be summarized as TCP

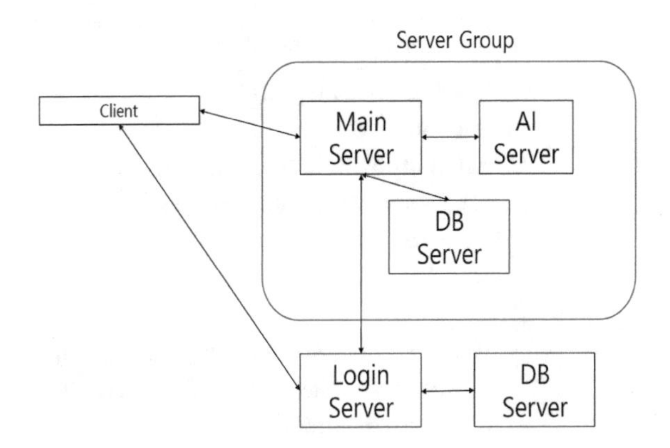

Fig. 3 Commercial game server architecture

simply passing a stream of bytes. How to handle the flow of unbounded bytes as much as the data sent by the user is explained. There are two ways to create a packet on a server using the C++ language. The first is to use Array in Fig. 4 and the second is to use structures and unions. There is no difference between the two methods. The structured design has the advantages of good readability, but it has the disadvantage of being dependent on C language and converting it when communicating between different programming languages. Conversion work is cumbersome and error-prone during the development process, so it is a prudent choice when communicating between languages.

Most MMORPG games using IOCP use TCP. In TCP, the order is guaranteed [4]. However, as mentioned above, there is no boundary between packets because it is a stream of bytes. The first part of the packet specifies the length of the packet, and the second reason is that there is no boundary between the streams mentioned above.

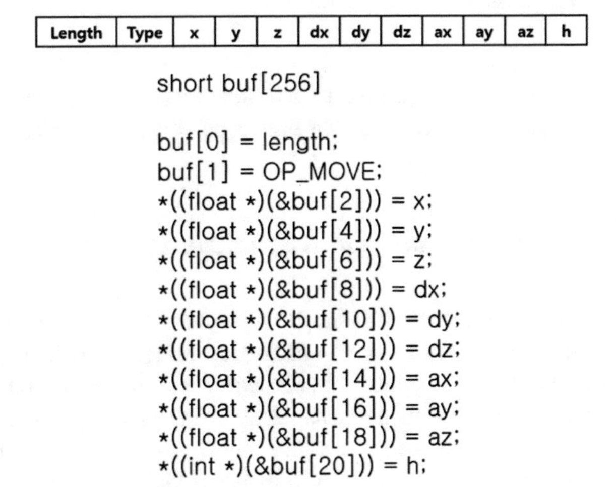

Fig. 4 Packet structure using array

Packet Assembly Algorithm

```
void Packet_Assembly(id) {
int rest = io_size;
CHAR *buf  = overlapped->io_buf;
packet_size = overlapped->curr_packet_size;
old_received = overlapped->stored_packet_size;
while (0 != rest)
{
    if (0 == packet_size)packet_size = buf[0];
        required = packet_size-old_received;
    if (rest >= required) {
        memcpy(overlapped->packet_buf+old_received, buf, required);
        ProcessPacket (id,overlapped->packet_buf);
        packet_size= old_received = 0;
        buf += required;
        rest -= required;
    }
    else {
        memcpy(overlapped->packet_buf+old_received, buf, rest);
        old_received+= rest;
        rest = 0;
    }
}
overlapped -> curr_packet_size= packet_size;
overlapped -> stored_packet_size= old_received;
```

Fig. 5 Packet assembly algorithm

Figure 5 is an algorithm to assemble incoming packets into each packet. When the packet is assembled, the process is processed through the ProcessPacket function.

2.3 *Optimization and Synchronization*

The result of the data race is not a problem with a single thread [5], but it can be seen from Table 1 that an unintended value occurs in a multithreaded environment. Figure 6 shows codes that can cause the data race. Table 1 shows the reason for the data race. This happens because one or more threads simultaneously read and write the shared variable Sum as shown in Fig. 6. In order to eliminate the data race phenomenon, it can be solved by using lock/unlock of the synchronization object in

Table 1 Results for various numbers of threads

Number of threads	1	2	4	8	16
Results	10,00,00,000	5,22,32,532	3,15,36,756	1,75,79,488	71,33,668

Fig. 6 Sum operation with multithread causing data race

Packet Assembly Algorithm

```
sum
function Thread_func(thread_num)
{
   for i from 1 to 100000000/thread_num
      sum += 1
   end for
}
```

the critical region [7]. In [8], we compared the performance of synchronized objects when using two threads.

In this paper, we propose a visual processing technique and a dead reckoning [9] technique for MMO game server optimization. In the server, when the information is transmitted to all users when broadcasting, a considerable overhead occurs in the server. In the case of the MMO game server, a packet is required if the connection is N. As a solution for reducing the overhead, a method of broadcasting only object information in the vicinity can be solved. All game characters have the concept of sight. The game client also improves its performance by curling the frustum itself, and the server introduces the concept of view of the game character similarly and sends only the object information around the game character, which helps to improve the performance of the server and reduce the amount of network transmission.

Because of the large size of the MMO game, it is essential to efficiently search for nearby objects. A server that has nothing to do with the client has introduced the Sector concept that divides the map into its own [10], so that only the objects near the character can be efficiently searched by searching only the sectors near the player. If the sector is too large, many objects outside the scope of the search are searched, and because the threads share one sector in the search process, the parallelism is reduced.

If the sector is too small, there will be frequent sector changes when moving the character, and overhead will occur because the object list of the sector needs to be frequently updated. Dead reckoning is an obstacle to sending and receiving large amounts of data due to the existing network problems like bandwidth and delay. In this case, the client may appear to be disconnected from the screen.

In fact, it is a technique to make a delay happen but not to feel such a situation. Dead reckoning in the game is mainly used to update the user's location information. The most basic player movement method is to broadcast a mobile packet based on the user's key input, which causes a huge server load on a large-scale online game server due to its large load. Assuming a 10-s move in a 60-frame game, a total of 600 packets will be sent to the server. If there are ten players around, you have to broadcast 6600 packets including that user.

The dead reckoning technique broadcasts a packet to move through the user's first keystroke. There is no packet from the client until another keystroke arrives, and the server only sends the updated location periodically. This period can be determined by the status of the server or the nature of the game. The client can also update the location by comparing the position of the character and the position of the packet periodically coming from the server.

Dead reckoning in game development, but the words of a very narrow meaning referring to the situation that developed so difficult to progress in the game the error originated as screen inconsistencies among users due to the time delay now means the synchronization techniques and predictive algorithms to eliminate it [11].

The time delay is the time it takes a packet to travel from the client to the server and back to the client. Due to the time delay, users will see different screens, which disrupts the fairness of the game. Developers use dead reckoning to keep users from experiencing errors due to time delays and to maintain game fairness. In a large online game, a server processes hundreds or thousands of clients, and if the clients send frequent status update packets to the object, the server is overloaded with network loads. Therefore, if a large-scale online games, including the time delay, also resolved a letter to reduce the amount of packet [12].

There are three ways to update the state. The first is a method in which the client and the server periodically transmit state variables to each other. The second is a method of transmitting the state update packet only when the state changes to a predetermined value or more. The third is to send the update packet even when the status changes to more than a certain value for an instant response, by sending the update packet at regular intervals. Three methods are all playing the game according to the predictive algorithm based on the last received status information, if not received the update packet [9, 13].

The authors are concerned with a card game called Daihinmin (Extreme Needy), which is a multi-player imperfect information game [14]. The UEC Computer Daihinmin Championship is held at The University of Electro-Communications every year, to bring together competitive client programs that correspond to players of Daihinmin, and contest their strengths. Wakatsuki et al. [14] extracts the behavior of client programs from actual competition records of the computer Daihinmin, and propose a method of building a system that determines the parameters of Daihinmin agencies by machine learning.

Comparing the wireless sensor network with other conventional network system is a tough job [15]. Therefore, since last decade the effort has been made to design and introduce a large number of a communication protocol for WSN with given concern on the performance parameter of energy efficiency and still the key requirements within WSN domain, that how to incrementally expands the energy minimization consuming techniques of sensor battery [15]. The other parameters include latency, fairness, throughput and delivery ratio. Haqbeen et al. [15] proposes a novel joint cooperative routing, medium access control (MAC) and physical layer protocol with traffic differentiation based QoS-aware for wireless sensor network (WSNs).

3 Implementation of a MMO Game Server

As following those in Fig. 3, this paper creates a simple MMO game server. For the use of libraries such as threads and synchronization objects, the language used was C++ 11, and the compiler used Microsoft Visual Studio 2015 [8]. The client also used the same compiler, and the graphics library used OpenGL [10]. The chess pieces are represented by moving chess pieces on the chessboard. The game server has eight threads excluding the main thread.

Packet processing and NPC logic processing, six worker threads for I/O, timer threads for continuously moving NPCs, and accesses threads for connecting users. The map was made at 500 × 500, and the size of flare and NPC was set to 1 × 1. We set the maximum number of users to 20,000 and put 5000 NPCs on the map.

When the server is first run, the NPCs are initialized. NPCs are randomly placed on the map, and the event to be acted on is inserted into the priority queue, which acts on the timer thread, with the id value, the time to act, and the action event. This priority queue determines the priority by comparing the time value for the NPC to perform the event.

Take the header part to the priority queue and let the WorkerThread process the NPC's logic through PQCS (PostQueueCompletionStatus). In the server, NPCs are set to move randomly every second. Table 2 shows the types of packets to be used.

Packets from the client to the server are packets indicating the direction through the key input. The server updates the direction in the worker thread according to the packet type through the ProcessPacket function, creates a packet in the SC_POS type, and broadcasts the location to the users in the field. The packet from the server to the client consists of SC_POS to inform the location, SC_PUT to let the player and NPC know when it is in sight, and SC_REMOVE to notify when the NPC or player is disappeared or disconnected. A visual processing technique was applied to the optimization technique used in the server. Dead reckoning is excluded due to the nature of the chessboard client moving one space per keystroke. Figure 7 is a simplified version of the structure of the game server.

Table 2 Packet type

Server -> Client	Client -> Server
SC_POS	CS_UP
SC_PUT	CS_DOWN
SC_REMOVE	CS_RIGHT
	CS_LEFT

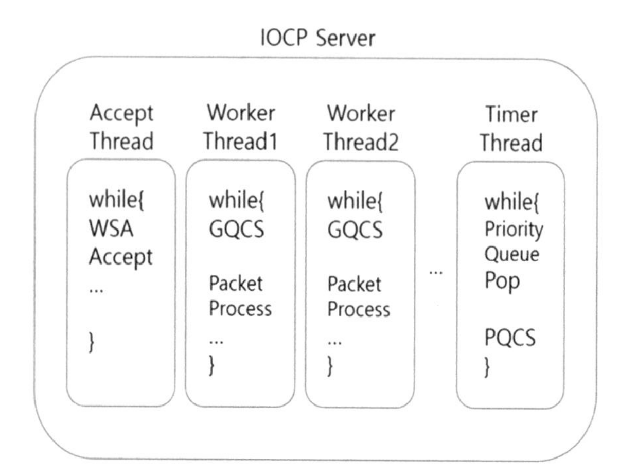

Fig. 7 IOCP game server architecture

4 Performance Evaluation Using Stress Test

In this paper, we performed a performance test using a stress test program. Experimental items were an overload experiment in which dummy avatars were concentrated at a hot spot and a maximum dummy experiment in which dummy avatars were evenly scattered. The performance of the computer used in the experiment is as follows: Processor: AMD Ryzen 7 1700 Eight-Coire Processor 3.00 GHz, Memory 16.00 Gb, Operating System Windows 10 Pro 64 bit Operating System Graphics Card: Geforce Nvidia GTX1060 6 GB.

Stress test program was also produced through IOCP like the server created in Sect. 3. DrawModule that can grasp real-time location of connected dummy avatars and NetworkModule to communicate with the server in real time. Network module connected dummy avatar continuously for a specified maximum number of users and based on connected avatars. The key input packet is sent to the server every second.

4.1 Hotspot Simulation

All the connecting clients are designated as coordinate systems 10 and 10 and then connected. In Fig. 8a, the part where the central point is gathered is around 10, 10 which is set as the hot spot point. When the center of the stress test program window is set to 0, 0 and the 500 × 500 map is reduced to 1/400 scale, it is possible to check the position of the whole dummy avatars at a glance in real time.

After connecting about 900 dummies, the connection was slowed down due to the load in the test program. Over time, the number of connections slowed down and the CPU utilization rises. Results of about ten tests are shown in Table 3. An average of 8721 dummy avatars received connections and did not receive connections from the server. After an average of about 8000 dummy avatars was connected, the

Fig. 8 Stress test

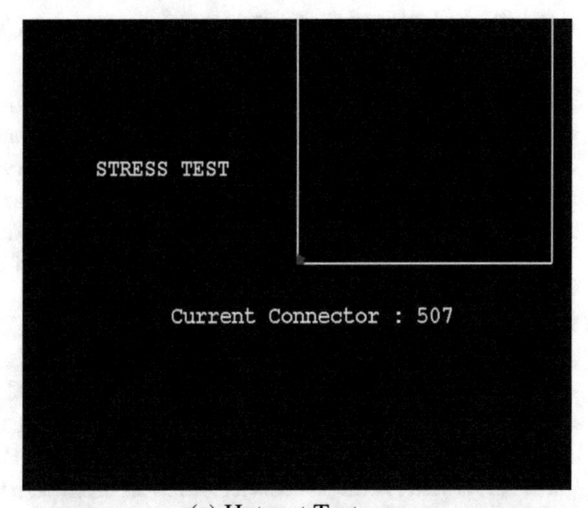

(a) Hotspot Test

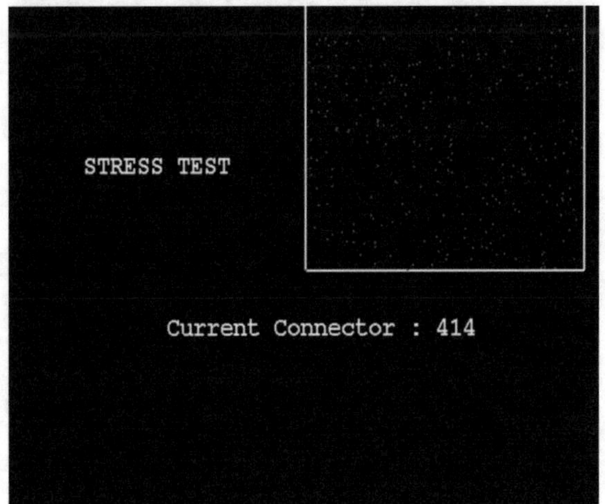

(b) Maximum connection Test

Table 3 HotSpot test result using stress test

Test number	1	2	3	4	5
Connections	8743	8732	8674	8903	8873
Test number	6	7	8	9	10
Connections	8594	8602	8876	8562	8657

CPU utilization was approaching 100%, and after the 8500 was exceeded, the time to connect increased by a factor of ten compared to just after the program was turned on.

4.2 Simulation of Maximum Concurrent Connection

The connecting dummy avatars were placed randomly throughout the map. As shown in Fig. 8b, it is possible to confirm that the dummy avatars are dispersed in the map evenly. Due to the nature of broadcasting to N^2, the servers were overloaded quickly at the hotspot points, while the load of the servers was relatively late, so we could connect a larger amount of dummy avatars than hotspot tests.

This server could connect a much larger amount of dummy avatars than hotspot tests. On average, about 19,324 dummy avatars were connected, and we could see that the server overload was maximized as shown in Table 4.

Besides, we can confirm that there is little difference between the maximum and minimum samples compared to the hotspot test. This MMO game server was able to get a lot more connections than being expected while testing. It is expected that there is no complicated logic, and the movement of the NPC is very simple. Perhaps a game server that is built for service purposes is expected to connect far fewer avatars if simulated with the same specifications as the computer specifications tested in this paper.

4.3 Simulation of Maximum Concurrent Connection Except Visual Field Processing

In the maximum tangent test conducted in Sect. 4.2, the simulation was performed except for the visual processing technique. The results are shown in Table 5.

Except for the visual processing technique, approximately 6000 dummy avatars were not able to connect to the result in Sect. 4.2. It was part of the game server to see how important optimization was. In addition, various optimization techniques are expected to connect more dummy avatars.

Table 4 Maximum concurrent connection test result using stress test

Test number	1	2	3	4	5
Connections	19,305	19,438	19,312	19,462	19,421
Test number	6	7	8	9	10
Connections	19,364	19,201	19,206	19,218	19,354

Table 5 Maximum concurrent connection test result using stress test where visual field processing is excepted

Test number	1	2	3	4	5
Connections	12,817	12,938	13,031	12,962	13,011
Test number	6	7	8	9	10
Connections	13,014	13,031	12,986	12,998	13,054

5 Conclusion

In this paper, we propose the basic structure for MMO game server and suggest synchronization scheme and packet structure for performance enhancement. The techniques proposed are data race using synchronization objects, packet loss reduction using deadlock prevention, field processing, and dead reckoning. In the MMO game server where the number of broadcasting increases in proportion to the number of server accesses, the above-described view processing technique and dead reckoning technique are essential. Experimental results show that the performance improvement of the game server can be expected only by using the above two techniques. The experiments above were conducted with LocalHost, so it is hard to trust absolutely on real commercial servers. Actual commercial servers will have more unpredictable parameters, and packet reception from clients will be more irregular than from experiments with stress tests.

Future research will focus on the use of multiple computers instead of a single computer to improve reliability in this paper.

Acknowledgements This work was supported by Institute for Information and communications Technology Promotion (IITP) grant funded by the Korea government (MSIP) (No. 2016-0-00204, Development of mobile GPU hardware for photo-realistic real time virtual reality).

References

1. Fall, K.R., Stevens, W.R.: TCP/IP Illustrated, Volume 1: The Protocols. Addison-Wesley (2011)
2. Loe, C.H., Seo, C.S., Wook, B.: Data priority-inheritance algorithm for deadlock prevention in distributed systems. In: Fall Conference of Korea Multimedia Society, pp. 106–111 (1998)
3. Choi, S., Park, H.-Y.: Study on the online game server architecture. In: Spring Conference of Korea Academia-Industrial Cooperation Society, pp. 534–538 (2006)
4. Jang, S.-M., Yo, J.-S.: An efficient MMORPG distributed game server. J. Korea Contents Assoc. **7**(1), 32–39 (2007)
5. Lee, N.-J., Gwak, H.-S.: The distributed server model for the evolutionary online RPG. J. Korea Game Soc. **2**(1), 36–41 (2002)
6. Moon, S.-W., Cho, H.-J.: A study on synchronization distribution of server message in online games. J. Korea Game Soc. **9**(2), 105–113 (2009)
7. Savage, S., Burrows, M., Nelson, G., Sobalvarro, P., Anderson, T.: A dynamic data race detector for multithreaded programs. ACM Trans. Comput. Syst. **15**(5), 391–411 (1997)
8. Aggarwal, S., Banavar, H., Khandewal, A., Mukherjee, S., Rangrajan, S.: Accuracy in dead-reckoning based distributed multi-player games. In: Proceedings of 3rd ACM SIGCOMM Workshop on Network and system support for games, NetGames '04, pp. 161–165 (2004)
9. Shim, K.-H., Kim, J.-S.: A study on performance analysis and improvement of dead-reckoning algorithm in networked virtual environment. Fall Conf. Korean Inst. Inf. Sci. Eng. **28**(2), 112–114 (2001)
10. Yu, S.-J.: Game server and spatial partitioning for MMORPG. Commun. Korean Inst. Inf. Sci. Eng. **23**(6), 29–35 (2005)
11. Kim, S.-R., Yun, N.-K., Koo, Y.-W.: Design and implementation of dead reckoning algorithm for network game. Korea Inf. Process. Soc. **7**(8) (2005)
12. Kim, K.-C.: Online game server. In: EGO, pp. 141–158 (2012)

13. Lengyel, E.: Believable dead reckoning for networked games. In: Game Engine Gems 2, pp. 307–327. A K Peters (2011)
14. Wakatsuki, M., Fujimura, M., Nishino, T.: A decision making method based on society of mind theory in multi-player imperfect information games. Int. J. Softw. Innov. (IJSI) **4**(2), 58–70 (2016)
15. Haqbeen, J.A., et al.: Design of joint cooperative routing, MAC and physical layer with QoS-aware traffic-based scheduling for wireless sensor networks. Int. J. Netw. Distrib. Comput. **5**(3), 164–175 (2017)

Automatic Generation of Image Identifiers Based on Luminance and Parallel Processing

Je-Ho Park, Young B. Park and Mi-Eun Ko

Abstract Recently, as the functionality of digital image acquisition devices is being improved, the commercial products with the high-performance functionality, such as smart phones and digital cameras, are being considered and utilized as an everyday commodity. As a result, it is natural that the volume of images which are collected from various paths or applications is also being enormously increased. According to this trend, the service platforms and applications, which support the functionalities necessary for image manipulation such as production, archiving and search, naturally demand the efficient management of enormous images either in independent or distributed systems. In such image management systems, an image identifier plays an important role in terms of identification of a particular image. Previous researches have been resolving problems either by applying complex methods that consume quite so much resources or by simple heuristic methods with the potential risk in terms of correspondence problem between identifiers and images. Therefore, the development of efficient and effective methods for the problem needs to be studied. In this paper, we propose a method to construct indexing of images utilizing the concept of the luminance area. The experimental evaluation of the proposed method illustrates that the proposed method satisfies the requirements for the image identification while reducing the processing cost.

Keywords Image identifier · Luminance · Parallel processing

J.-H. Park · Y. B. Park (✉)
Department of Software Science, Dankook University, Juk-Jeon, Yongin, South Korea
e-mail: ybpark@dankook.ac.kr

J.-H. Park
e-mail: dk_jhpark@dankook.ac.kr

M.-E. Ko
School of Computer Engineering, Hansung University, 116 Samseongyoro-16gil Seongbuk-gu, Seoul, South Korea
e-mail: mieun0518@gmail.com

R. Lee (ed.), *Software Engineering Research, Management and Applications*, Studies in Computational Intelligence 789, https://doi.org/10.1007/978-3-319-98881-8_3

1 Introduction

Recently, as the digital image acquisition functionality is being improved, the components or devices with the high-performance functionality such as smart phones and digital cameras have been quite expanded. Due to the popularization of such devices in various application areas, the volume of images which are collected from various paths or applications is also being enormously increased. As the produced images from the devices are being increased exponentially, the image and video based applications naturally demand the efficient management of enormous images either in the independent or distributed systems. According to a report by Cisco Visual Networking Index [1], worldwide connectivity of smart devices with multimedia and computing capabilities in 2016 accounted for 46% of all mobile device connections and 89% of mobile data traffic. Also, according to the report by Pew Research Center [2], 54% of Internet users are posting their original photos or videos online, and 36% of Internet users post their pictures or videos to share with others. Taking the above-mentioned statistical research results into consideration, enormous amount of image data is currently being generated and posted on the Internet applications. Regarding this situation, it is necessary that a system with effective and efficient image management capacity needs to be studied. In this paper, we concentrate upon a method that can uniquely identifies an image among a voluminous image set providing the effective image storage and retrieval capacity.

For a large volume of images, an identification method plays an important role in terms of identification of a particular image. In generally being used systems, in order to identify or distinguish an image from the other images, an arbitrarily assigned or automatically generated by a system identifier or a file name is being used to satisfy the identification requirement. However, these primitive methods can cause problems such as file name conflicts or duplicate identifier assignments with a recycling manner. Many studies [3–5] have been carried out to resolve this indiscriminate identifier problem. However, previously studied methods have to consider a large number of coefficients and to consumes high cost in terms of processing time.

In this paper, we propose a method to construct multiple indexing based on the luminance area. Our experimental results by comparing the proposed method with the existing methods illustrate that the proposed method satisfies identification requirement while reducing the processing cost by manipulating information flow.

2 Background for Image Identifier Generation

Generally, in a system that needs to manipulate the images for some particular objectives, it might uses a method that of automatically generates an identifier or a method that is arbitrarily given by a user to identify an image. However, a potential problem exists in such primitive methods. For example, when a number of distributed image management systems are integrated into a total system, the file name collision for the same image or duplicated identifiers for the different images might occur resulting in breaking the 1-to-1 correspondence requirement between images and identifiers. Many studies have been conducted to solve this indiscriminate identifier problem. A typical example is the method using image feature information [3–5]. Another potential problem is that the processing time is relatively high. As a method for resolving the problem of identifier generation, a method constructs an image identifier based on a luminance area [6]. The method utilizes the luminance related information of an image and the normalization of the collected information to produce an image identifier. In order to provide the insight of the image identifier generation, here we provide the background of the concept and the essential ideas for the image identifier generation.

2.1 Line Component Based Approach Generation

The line component based scheme utilizes the edges or the line segments which can be detected from an image. In this scheme, first, an input color image needs to be converted into a grayscale image. And then, an edge detection algorithm is applied to the converted grayscale image. The detected edges are used for line segment composition which is a sequence of edges. In Fig. 1, we demonstrate the sequence of the process for an image identifier generation. The first image in Fig. 1 shows the input image. After detecting edges in the image by using an edge detection algorithm such as Canny algorithm [7], the statistical Hough transformation [8] is applied to the image in order to extract the line segments. In the second image of Fig. 1, the edges detected by an edge detection algorithm are illustrated in gray, while the line segments are shown in red. In (d) of Fig. 1, the part of the image with the line segments is shown in order to visualize the composed line segments in detail.

The essential idea supporting the first approach is the random distribution of edges and line segments in an image. In order to describe our approach, *LinearSeq*, which is the set of pixels, p, in the line segments, is defined as follows:

$$\mathrm{LinearSeq} = \{\mathrm{p(x, y)} | \mathrm{p(x, y)} \in \mathrm{LineSeg}\} \tag{1}$$

The relationship between the pixels in LinearSeq and the set of virtual paths, *VPath*, which are predetermined by our approach can be described as follows:

$$\mathrm{CPoints_l} = \{\mathrm{p(x, y)} | \mathrm{p(x, y)} \in \mathrm{LineSeg} \wedge \mathrm{p(x, y)} \in \mathrm{VPath}\} \tag{2}$$

The predetermined virtual lines over the image under processing are used for an efficient quantification process. The virtual lines are used in order to collect the intersecting locations. Moreover, this collected information, the frequency of intersections, is used for quantization for our image identifier composition as a constituent value for an identifier like following:

Fig. 1 The effect of edge and line segment detection (**a** original, **b** edges, **c** line seg., **d** edges and line segments in detail)

$$\mathrm{ID_{img}} = \{(\mathrm{i_0} \ldots \mathrm{i_v})|\mathrm{i_j} = \mathrm{Card(CPoints_l)}, \quad \text{for all j} \in \text{virtual lines}\} \tag{3}$$

The virtual lines that spread out from the center of an image and are extended to the boundary of the given image. In order to increase the size of an image identifier, which creates more accuracy, the more virtual lines can be used on demand. The middle points of two adjacent points on the boundary of an image are regarded as the end points for additional virtual lines from the center. Figure 2 illustrates an example with eight virtual lines.

2.2 *Luminance Area Based Approach*

With the line component based approach, the extended line segments created by Hough transformation provide another information in addition to the edges. However, if large intervals between line segments appears frequently, the collected information with relatively less randomness does not sustain the successful identifier creation. This undesirable situation can be recognized in the dark part of the image shown in Fig. 1b. In [6], we studied image identification using the luminance area and the segment based image identifier generation methods. A luminance area is defined as

Fig. 2 Eight virtual straite lines

a set of adjacent pixels with the same luminance value. Using the luminance area concept, we view an image as a collection of luminance areas.

If we decrease the degree of luminance levels, the adjacent luminance areas are merged into a larger luminance area. This is the essential feature of our second approach. If the luminance degree is controlled in well-organized manner, the collection of information regarding the distribution of luminance areas can be excellent guide for an efficient identifier generation. If we decrease the degree of luminance levels, the adjacent luminance areas are merged into a larger luminance area. This is the essential feature of our second approach. If the luminance degree is controlled in well-organized manner, the collection of information regarding the distribution of luminance areas can be excellent guide for an efficient identifier generation.

For the description of our second approach, we describe a luminance area like as follows:

$$\mathrm{LArea} = \{\mathrm{P_i} | \mathrm{L}(\mathrm{P_i}) = \mathrm{L}(\mathrm{P_j}) \text{ and } \mathrm{P_i} \text{ is transitively adjacent to } \mathrm{P_j}, \quad \text{for all } \mathrm{P_j} \in \mathrm{LArea}\} \tag{4}$$

where L describes the luminance degree of the pixel P_i. Following the definition, every pixel in a luminance area is transitively adjacent. Figure 3 shows grayscale images with varying luminance scale with 256–4 levels respectively. The luminance degree for the general images has 256 levels. The right-bottom image with 4 luminance levels can be perceived as very unnatural due to human eyes, however, this process is not for the image enhancement but for the generation of an image identification. Unlikely the distributed line components, the luminance areas partition an image explicitly even within the image regions in which line component composition fails.

The distribution of luminance areas is quantized by using the virtual lines. The boundary information between areas in which there exists discrepancy between two areas regarding luminance is collected for the generation of an image identification. Like the line component based approach, the luminance area based approach can be implemented in a similar manner by processing transformation over the whole image area. However, in the sampling process of areas, the mapping among different lumi-

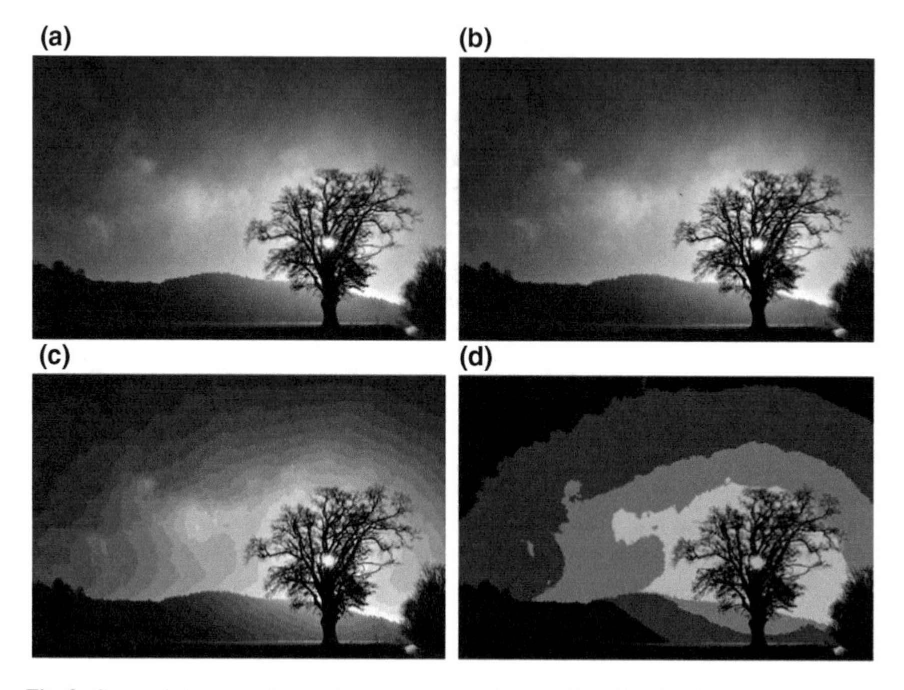

Fig. 3 Grayscale images with varying luminance scales (**a** 256, **b** 64, **c** 16, **d** 4)

nance scales can be performed only in necessary pixels along with the virtual lines. This results in a huge saving in processing time comparing to the line component based method. These two image identifier generation approaches can be resolved within the divide-and-conquer pattern. In order to discover another enhancement opportunity in the context of image identifier generation, in the following, we apply parallel processing technique to our interesting study.

2.3 Parallel Processing Based Approach

Apache Hadoop is an open source framework that provides an operating foundation for distributed processing and storage of large amounts of data. It is designed to expand from a single server to multiple servers, and provides a distributed file system capable of storing large amounts of data and a parallel processing system capable of processing large amounts of data. Moreover, Apache Hadoop is Java-based and provides a Hadoop Map-Reduce programming model. This Hadoop Map-Reduce model is based on the Hadoop Distributed File System (HDFS) and provides a Java-based API to enable the creation of parallel processing applications [9, 10].

OpenCV is an open source computer vision library that can be used on multiple platforms. HIPI is a library designed to handle images in the Hadoop Map-Reduce model and provides integration with OpenCV. In this context, if the integrated environment is established between the two libraries, the image processing techniques can be utilized in the Apache Hadoop environment [11, 12].

In this regard, if an environment integrating a parallel processing operation base for efficiently storing and processing a large amount of data and a library for supplementing the parallel processing operation is constructed, efficient processing will be possible. Hadoop corresponds to a parallel processing operation base, and HIPI and OpenCV correspond to a library for image processing in a parallel processing environment. Figure 4 shows the operation structure of the integrated environment.

The processing flow is divided into three phases: input phase, cull phase, and Map-Reduce phase. The input phase is the step of entering the locally existing image into Hadoop HDFS. The HIPI library is used when inputting to Hadoop HDFS. HIPI provides an image bundle called Hipi Image Bundle (HIB) [13] for efficient processing of the large images. To generate these HIB files, HIPI provides a tool called hibImport [14]. The data input phase transfers large-capacity images from the local file system to HDFS. To do this, the hibImport tool to generate the HIB file is utilized. And, to save the created HIB file in HDFS, it executes the process of inputting the command to the terminal. The following is the cull phase. The cull phase is to extract an image from the HIB file at runtime in the Map-Reduce program, and performs image culling [15] before passing it to the Mapper. Finally, it is the Map-Reduce phase. Map-Reduce [10] is created through the implementation of the abstract classes Mapper and Reducer. Map-Reduce processing is performed in three steps: Map, Shuffle, and Reduce. Each step has a key-value pair as input and output. First, Map is an individual task that converts input records into intermediate records, and sets the key. The second is Shuffle, which is used to sort the output of each Map task by keys and input them to the Reduce task. Finally, the Reduce step performs key-specific data processing on the sorted data. Map-Reduce phase is implemented by this Map-Reduce program. Here, HIPI and OpenCV provide functions for image processing.

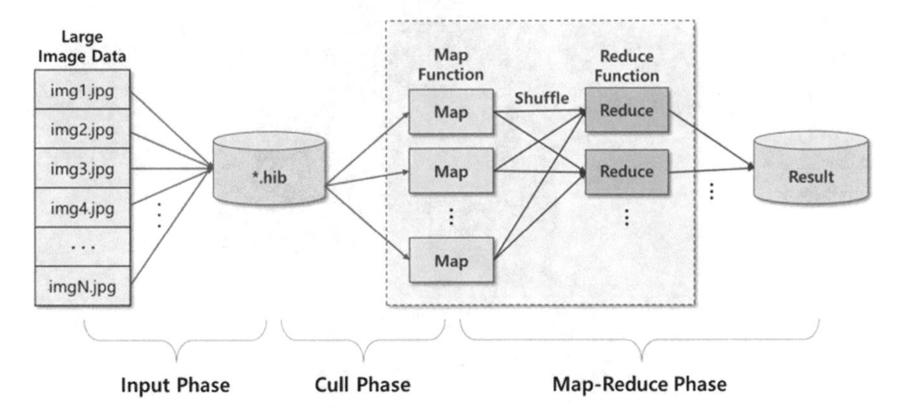

Fig. 4 Operation structure of parallel processing based image processing

2.4 Multiple Index Generation Based on Luminance Area

In this paper, we propose a multiple index generation method based on luminance domain. The proposed method applies the divide-and-conquer pattern to both the luminance area based approach and the segment based image identifier generation method. And we apply parallel processing technology to improve performance. The proposed method uses the Hadoop Map-Reduce programming model, HIPI and OpenCV. The overall workflow is shown in Fig. 5.

The overall process shown in Fig. 5 performs the following four phases: data input preparation phase, data input phase, linear element expansion and filtering phase, regularization phase. First, explain the input preparation phase. This phase controls the preparation of the test data. At this point, the determination of a size is important so that the size of the test data is large enough to be managed by the Hadoop distributed file system. Second, explain the input phase. This phase is to provide the selected test data to the Hadoop distributed file system. In this phase, HibImport tool provided by HIPI used for input. As a result of the phase, HDFS is input with the extension ".hib". This file is the HIPI image bundle created by the HibImport tool. Third, explain the linear element expansion and filtering phase. This phase generates an identifier for the provided image in the HDFS. The phase has four steps: indexing setting, DDA algorithm, luminance formula, Rein-hard tone-mapping formula. Figure 6 Shows the processing flow in the linear element expansion and filtering phase in detail. A detailed explanation is given below.

At first, an 8-direction virtual line is set up for index generation, and two virtual linear operations are allocated in order to process the 8-direction virtual line. And

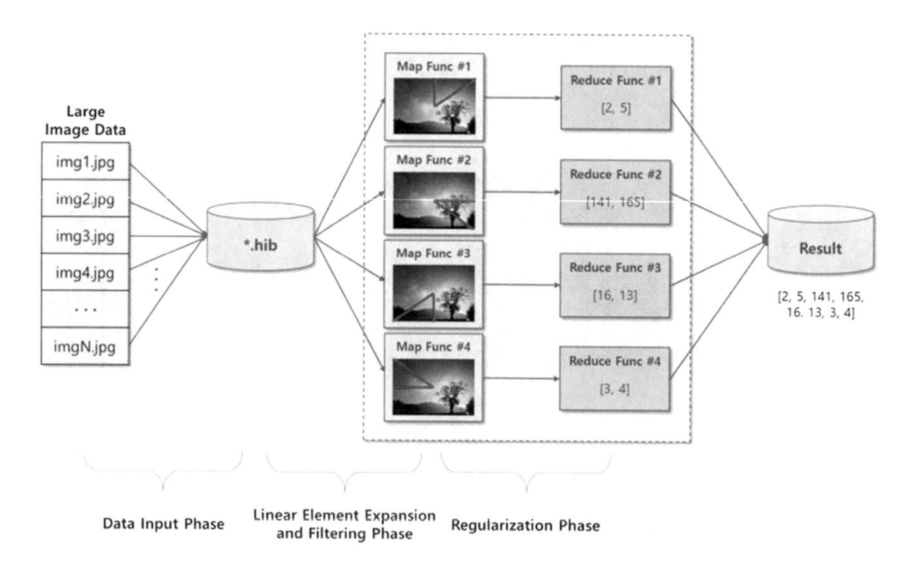

Fig. 5 Index creation structure

then, the DDA algorithm [16] is applied to the pixel near the virtual line as the reference point. The DDA algorithm is an algorithm that uses the slope to find the next point when finding the next point at the start point. Formula 1 is the slope formula.

$$M = (y_{end} - y_{start}) / (x_{end} - x_{start}) \tag{5}$$

Considerations for determining the position of the pixel are as follows. Formula 2 is to be considered when the slope is less than or equal to 1, and Formula 3 is to be considered when the slope is greater than 1.

$$\text{If } m <= 1,\ x_{k+1} = x_k + 1,\ y_{k+1} = y_k + m \tag{6}$$

$$\text{If } m > 1,\ x_{k+1} = x_k + 1/m,\ y_{k+1} = y_k + 1 \tag{7}$$

After applying the above algorithm, perform the following luminance formula and tone-mapping filter. And the result is reflected in the next step. The luminance is measured for filtering, and a tone-mapping filter is applied. Formula 6 is the relative luminance [17] formula for luminance measurement.

$$LRGB = 0.2126 * R + 0.7152 * G + 0.0722 * B \tag{8}$$

Equation 9 is the Rein-hard Tone-Mapping [18] formula for applying the tone-mapping filter. Finally, explain the regularization phase. The regularization phase generates identifiers using the results derived from the linear element expansion and filtering phase. The method is as follows. The time when the variation of the luminance occurs is counted, and the result is used as a part of the identifier.

$$L_{out} = L_{in} / (L_{in} + 1) \tag{9}$$

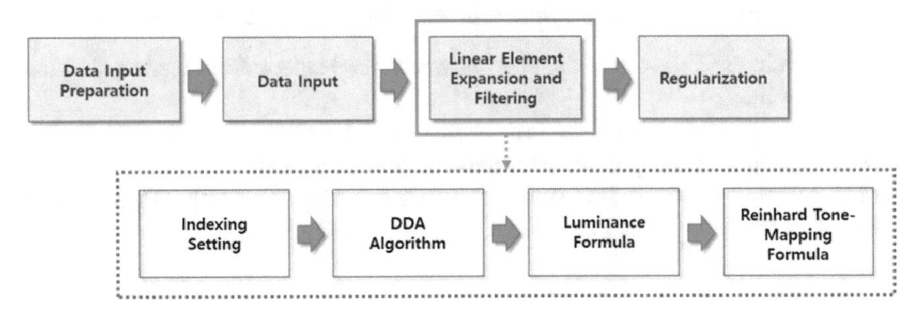

Fig. 6 Processing flow of the method of generating multiple indexing based on luminance area

3 Background for Experimental Evaluation

3.1 Experimental Environment

For performance evaluation, we ran experiments with Hadoop cluster configurations of one master and four slaves. The master node served as the Name-Node, Secondary Name-Node, Resource Manager and Job-History Server, and the four slave nodes served as the Data-Node and Node Manager. We used Hadoop, HIPI and OpenCV as tools for the experiment and used Linux environment as the operating system on which the tools were run. A detailed specification of the tools used in the experimental environment is as follows: Apache Hadoop 2.6.0, HIPI 2.0, OpenCV 2.4.11, Java 1.7, Ubuntu 14.04 LTS.

The following is a description of the input data environment. We used two types of input data sets. The input data set is divided into two types: Small-Size-Images dataset and Large-Size-Images dataset. The Small-Size-Images dataset was selected with 10,000 image data sizes less than 0.2 MB. And Large-Size-Images dataset was selected as 300 image data of 5 MB or more size. A detailed description of the input data used in the experiment is shown in Table 1.

3.2 Performance Metrics

This experiment compares and analyzes the data processing performance of the proposed method. In this context, we defined internal execution steps of the Hadoop system to analyze performance in various aspects. And we defined the cost that occurs in the execution process. First, it defines the internal execution steps of the Hadoop system. Hadoop is divided into five execution states: job submitted, map attempt done, map attempt finished, reduce attempt started, reduce attempt finished, and job finished. A detailed description of each execution status is shown below.

- Job Submitted: This step is the state in which the Hadoop Map-Reduce Job has been submitted.
- Map Attempt Started: This step is the state in which the initially assigned Map has been started. In this paper, we refer to the first started Map.
- Map Attempt Finished: This step is the state in which the finally assigned Map has been finished. In this paper, we refer to the last finished Map.

Table 1 Image samples size information

	Small sized images	Large sized images
Image count (images)	10,000	300
Average size (KB/image)	122	7521

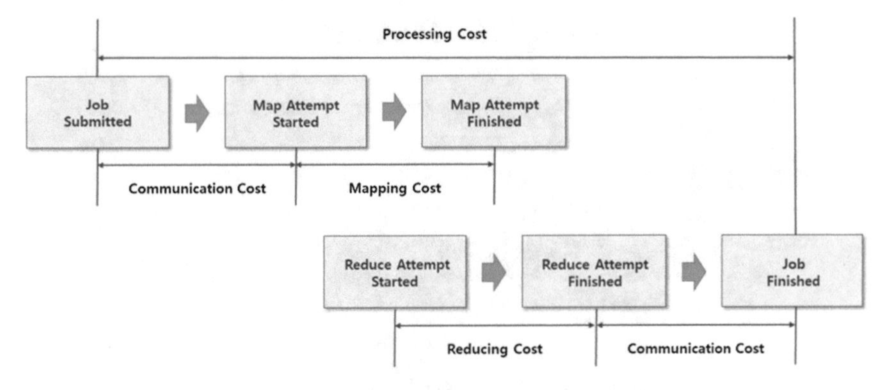

Fig. 7 Structure of Hadoop's internal execution step definition and structure of cost definition in execution

- Reduce Attempt Started: This step is the state in which the initially assigned Reduce has been started. In this paper, we refer to the first started Reduce.
- Reduce Attempt Finished: This step is the state in which the finally assigned Reduce has been finished. In this paper, we refer to the last finished Reduce.
- Job Finished: This step is the state in which the Hadoop Map-Reduce Job has been finished.

Second, it is the definition of cost that occurs during execution. Cost is divided into four cases: processing cost, communication cost, mapping cost, reducing cost. A detailed description of each cost is shown below.

- Processing Cost: This cost is the time it takes to perform a Hadoop Map-Reduce Job.
- Communication Cost: This cost is the time it takes to perform network I/O.
- Mapping Cost: This cost is the time it takes to perform Map.
- Reducing Cost: This cost is the time it takes to perform Reduce.

Figure 7 shows the internal execution steps of Hadoop and the time required for each step.

3.3 Performance Analysis

We compared and analyzed the performance results based on the criteria defined in the upper section. For the performance analysis, the linear coefficient based multiple indexing method which was previously studied was adopted as a comparative object. The results of the comparison object and the proposed method are shown in Figs. 8 and 9. The x-axis represents the type of cost, and the y-axis represents cost rate. And cost refers to the amount of processed images per millisecond, in KB/msec. In

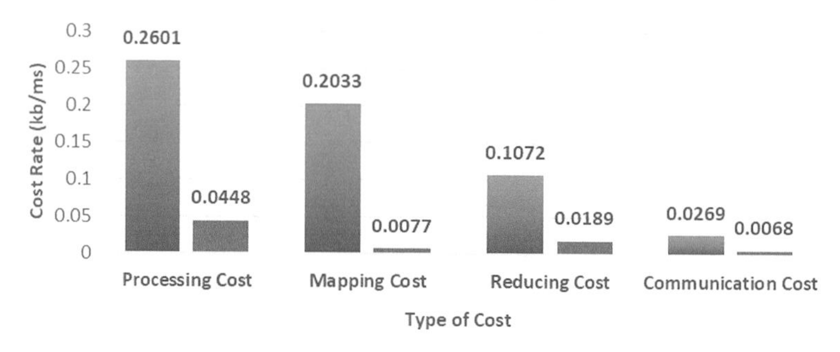

Fig. 8 The performance analysis results of the small size data set

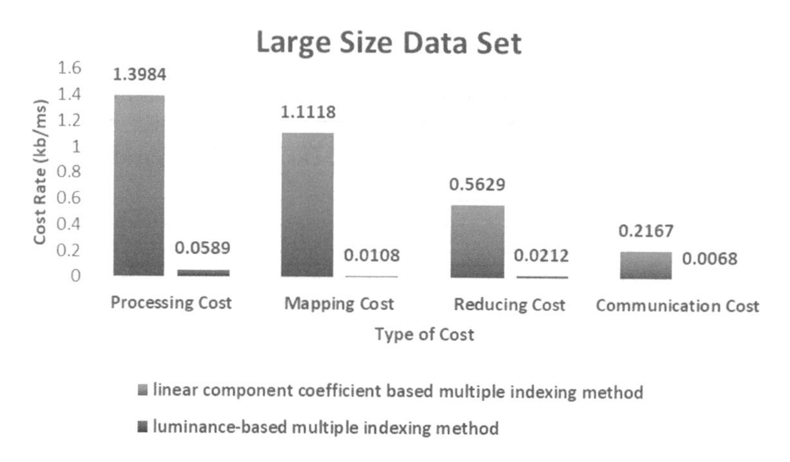

Fig. 9 The performance analysis results of the large size data set

addition, the blue bar represents the experimental results for the comparator, and the orange bar represents the experimental results for the proposed method.

The results of comparative analysis of Figs. 8 and 9 are as follows. Figure 8 shows the performance analysis results of the small size data set. The results show that the proposed method has lower cost than the comparison method in all cost cases. Figure 9 shows the performance analysis results of the large size data set. As with the previous results, the second result was obtained that all costs had less cost than the comparison method. Therefore, it can be seen that the proposed method has better overall performance than the comparison method.

In order to more accurately analyze the experimental results at the top, we analyzed the results with the improvement rate. The graph of improvement rate is shown in Fig. 10. In Fig. 10, the green bar represents a small set of data and the yellow bar

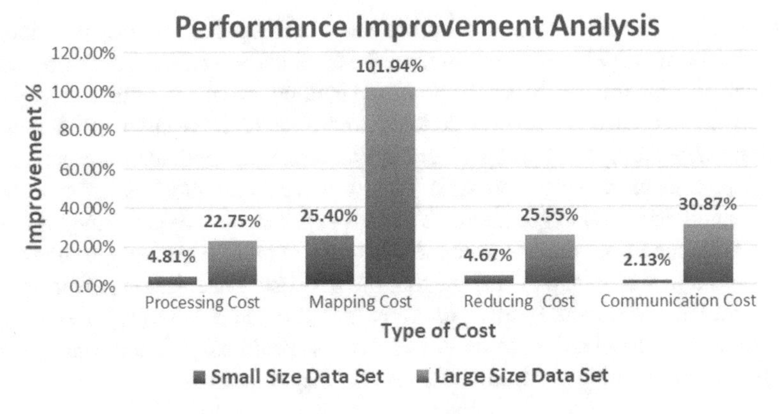

Fig. 10 The improved performance analysis results

represents the size of the experiment results for the proposed method. Now we are positioned to evaluate the performance of the proposed method through the derived improvement graph.

First, we compared the small size data set and the large size data set. The evaluation is as follows. All 4 cost cases were derived as high performance improvement rate in large size data set. Based on these results, the proposed method seems to be an advantageous method for large size images. The second is an evaluation of the results of the Small-Size-Set experiment. The mapping cost is the most improved in Small-Size-Set. This suggests that the proposed method is a concise method by reducing the number of steps compared to the existing method. Finally, it is the test result evaluation of Large-Data-Set. Again, the performance improvement of the mapping cost is obvious as in the second case. However, unlike the results of Small-Size-Set, the performance improvement of communication time is excellent. While the conventional method has the problem that the time taken to transmit the network from the large image deteriorates the overall performance improvement, the proposed method shows improved results.

4 Conclusion

Existing identifier generation methods have a problem of a large number of coefficients to be considered and problems of implementation complexity. Therefore, if an identifier of a large volume image is generated by an existing identifier generation method, a large amount of resources will be consumed. In order to solve this performance problem, a simple identifier generation algorithm is required, and a method of generating identifiers by parallel processing is needed. There is a luminance area-based indexing method with the existing identifier generation algorithm. In addition, there is a method for generating multiple indexes as an existing research for generating identifiers by parallel processing. In this paper, we propose a method of generating multiple indexing based on the luminance area for generating image identifiers in large images. Since the proposed method uses the luminance area-based indexing method, there are few coefficients to consider and it is easy to implement. In addition, since multiple indexing generation method is applied, identifier generation through parallel processing is possible. Finally, this paper implemented the proposed method and performed experiments. Through experiments, we confirmed the overall performance improvement over the existing research. Especially, experimental results of large size image data show that the network transmission performance is improved. However, there is still a problem that the time required for executing the algorithm is much occupied. In order to solve this problem, it is necessary to study efficient multi-indexing generation method in future research.

Acknowledgements This research was supported by The Leading Human Resource Training Program of Regional Neo industry through the National Research Foundation of Korea(NRF) funded by the Ministry of Science and ICT (No. NRF-2016H1D5A1909989).

References

1. Cisco Visual Networking Index: Global Mobile Data Traffic Forecast Update, online: https://www.cisco.com/c/en/us/solutions/collateral/service-provider/visual-networking-index-vni/mobile-white-paper-c11-520862.html
2. Photo and Video Sharing Grow Online, online: http://www.pewinternet.org/2013/10/28/photo-and-video-sharing-grow-online
3. Laaksonen, J., Koskela, M., Oja, E.: PicSOM—Self organizing image retrieval with MPEG-7 content descriptions. IEEE Trans. Neural Networks **13**, 841–853 (2002)
4. Datta, R., Joshi, D., Li, J., Wang, J.Z.: Image retrieval: ideas, influences, and trends of the new age. ACM Comput. Surv. **40**, 1–60 (2008)
5. Smeulders, A.W.M., Worring, M., Santini, S., Gupta, A., Jain, R.: Content-based image retrieval at the end of the early years. IEEE Trans. Pattern Anal. Mach. Intell. Arch. **22**, 1349–1380 (2000)
6. Pak, J.: Low-cost image indexing for massive database. Multimedia Tools Appl. **74**, 2237–2255 (2015)
7. Canny, J.: A computational approach to edge detection. IEEE Trans. Pattern Anal. Mach. Intell. **PAMI-8**, 679–698 (1986)
8. Kiryati, N., Eldar, Y., Bruckstein, A.M.: A probabilistic Hough transform. Pattern Recogn. **24**, 303–316 (1991)
9. Hadoop 2.6 hdfs, online: https://hadoop.apache.org/docs/r2.6.0/hadoop-project-dist/hadoop-hdfs/HdfsUserGuide.html
10. Apache Hadoop, online: https://hadoop.apache.org/docs/r1.2.1/mapred_tutorial.html#Prerequisites
11. HIPI: Hadoop Image Processing Interface, online: http://hipi.cs.virginia.edu
12. Widipedia, OpenCV, online: https://en.wikipedia.org/wiki/OpenCV
13. Class HipiImageBundle, online: http://hipi.cs.virginia.edu/javadoc/org/hipi/imagebundle/HipiImageBundle.html
14. Tool/hibImport, online: http://hipi.cs.virginia.edu/examples/hibImport.html
15. Class Culler, online: http://hipi.cs.virginia.edu/javadoc/org/hipi/mapreduce/Culler.html
16. Digital differential analyzer (graphics algorithm), online: https://en.wikipedia.org/wiki/Digital_differential_analyzer_(graphics_algorithm)

17. Relative luminance, online: https://en.wikipedia.org/wiki/Relative_luminance
18. Reinhard, E., Stark, M., Shirley, P., Ferwerda, J.: Photographic tone reproduction for digital images. ACM Trans. Graphics (TOG) **21**, 267–276 (2002)

Je-Ho Park He is a professor in the Computer Science and Engineering Department at Dankook University, South Korea. His research interests include multimedia database system, storage architecture, multimedia metadata, and indexing. Park has a Ph.D. in multi-layered database architectures from the Polytechnic Institute at New York University.

Young B. Park He received the M.S. and Ph.D. degree from the department of Computer Science, N.Y. Polytechnic University (NYU-Poly) in 1991. He is currently a professor in Dankook University. His research interests are in the areas of Intelligent Software Engineering, Automatic Software Testing, Software Development Process Enhancement and Software Refactoring.

Mi-Eun Ko She got a Master degree Computer Engineering, Dankook University, 2018. Now, she is working as an engineer in an IT Company. Her research interests include multimedia database system and medical imaging processing. Park has a Ph.D. in multi-layered database architectures from the Polytechnic Institute at New York University

Interface Module for Emulator-Based Web Application Execution Engine

Hyunwoo Nam and Neungsoo Park

Abstract The source code of a web application executed on the web browser has been exposed to leakage, and it is difficult to implement various web functions compared with other general applications. Therefore, in practical web applications, web plug-in technologies are used. However, web plug-in technology has a platform dependency and is attacked by malicious codes which resulting in a damage of the system. On the other hands, the web-based emulator is independent of platforms as well as web browsers. Recently, various web emulators were introduced, and users can use it as the virtual machine environment of diverse platforms on the web browser. However, these web-based emulator doesn't provide an interface API to access other web application. In this paper, we propose the modified web-based emulator with the interface module and API, called as the web emulator-based execution engine. Using the proposed system, a native code in a web application is easily written by C, and C++ and compiled. Furthermore it can be distributed and executed directly on the emulator-based web application execution engine. Therefore it solves the leak problem of source codes that is the vulnerability of web applications. The proposed emulator-based web application execution engine is a new concept combined web plug-in and emulator. In the experiment we developed an interface module and an API for the proposed emulator-based web application execution engine. The experimental emulator-based web application was implemented and tested to evaluate the overall system.

Keywords Web application · Obfuscation · Emulator · JavaScript
Native binary

H. Nam · N. Park (✉)
Dept of Computer Science and Engineering, Konkuk University, Seoul, Korea
e-mail: neungsoo@konkuk.ac.kr

H. Nam
e-mail: namhw@konkuk.ac.kr

R. Lee (ed.), *Software Engineering Research, Management and Applications*, Studies in Computational Intelligence 789, https://doi.org/10.1007/978-3-319-98881-8_4

1 Introduction

The latest web technologies have been used widely on the application of environment such as mobile, PC, server and IoT service as well as a web service. However, many companies have difficulties in protecting the source code of the web application serviced using only web standards technology. Because the source code, the enterprise asset in a web service, is downloaded and executed based on the script format in the user's web browser without additional encoding, the source code of the script-based web application is directly and easily exposed through the web browser developer tool, resulting in the leakage of the source code.

The practical approaches to protect the source code of a web application are to block the right mouse click viewing the web source code or to obfuscate the source code. However, even though applying these approaches, the source code can be still viewable from the Web browser developer tool. Also, it is getting difficult to protect the source code due to some tools which restore the obfuscated code.

As other approaches to protect the source code efficiently, the core logic in the source code of a web application is relocated to the server or is implemented using the web plug-in technologies such as ActiveX, NPAPI (Netscape Plug-in Application Programming Interface). It can prevent fundamentally the source code from being leaked because it can be executed on the server or distributed as an executable binary file. However, the server-based approach has a disadvantage of increasing the server load and network traffic. Also, web plug-in technologies are infected by other malicious code which leads not used on modern browsers.

In this paper, we propose the emulator-based web application execution engine that can use the executable binary file which is possible to prevent from leaking source codes and to perform the various functions on the web environment. Furthermore, the interface module and API are necessary for the emulator-based web application execution engine to communicate with other web applications. The proposed approach makes the executable binary for the web application available in the web standard environment without any web plug-in technology. In this paper, we designed and implemented the interface module and API in the emulator-base web application execution. In addition, an emulator-based web application was developed by using the developed interface API for experiments. In the experiment, the performance of the emulator-based web application is shown competitive on several platform and evaluated the source code protection of the proposed approach in various points of views.

The rest of a paper consists of the followings. In the next chapter, the background and the motivation of this research will be presented. In the chapter 3, the proposed emulator-based web application engine will be introduced and discussed. The chapter 4 will explain how to implement an emulated-based web application and its interfaces. The chapter 5 will show the experiment results and discuss the protection of a native code in a web application using the proposed approach. In the last chapter, the conclusion will be remarked.

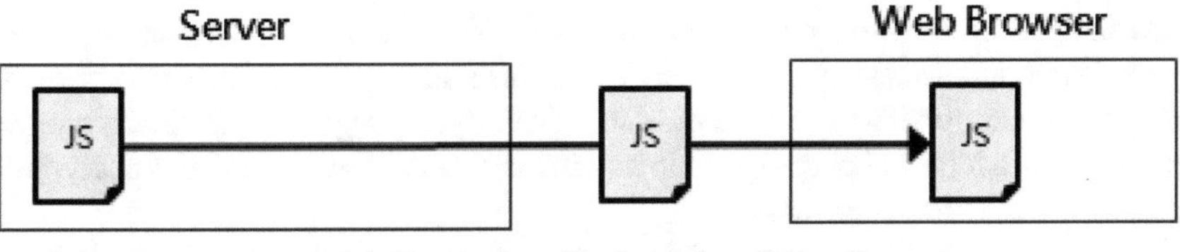

(a) General script-based web service

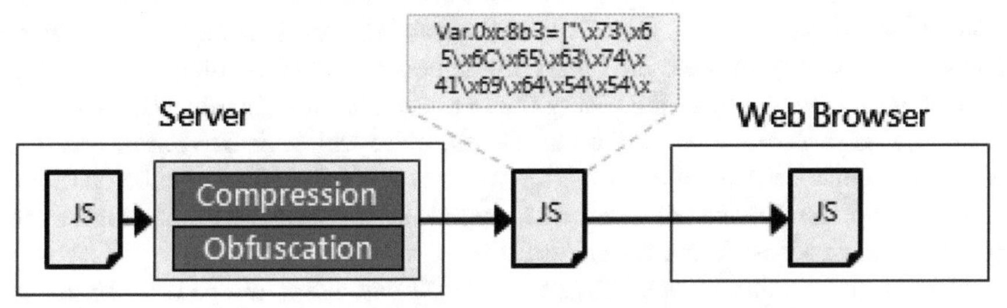

(b) Web service using an obfuscation approach

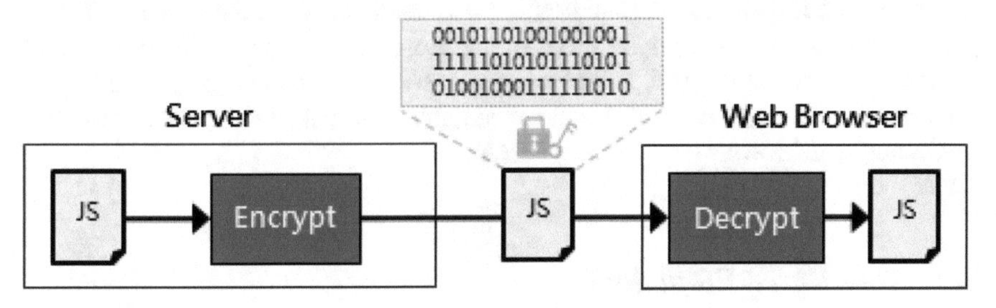

(c) Web service using a cryptographic approach

Fig. 1 Type of source code distributions for a web service

2 Background and Motivation

2.1 Limitation of Web Source Code Protection Technology

In the general script-based web application, its source code is possible to leakage easily because its original source code is distributed as the text-based script language such as Fig. 1a. One of well-known approaches is to hide the source codes of a web application, which is called as an obfuscation. The obfuscation of a source code is the conversion of the source code of an application into a structure that is difficult for humans to understand as shown in Fig. 1b. The obfuscated source code can be directly executed on the browser without additional decode process. However, it is possible to restored the obfuscated source through deobfuscating tools [1].

Another approach to protect the source code is to encrypt and package the source code of the web application [2]. Figure 1c shows a cryptographic deployment method that can be secured during the creation and distribution of a web application. However, it is still possible to leakage source codes because they can be decrypted before the encrypted source codes are executed. Also, other disadvantages are that the cryptosystem requires additional cryptographic key management scheme, and the decryption process is added in the execution.

In case the high protection of the source code is required, the web plugin technologies are used. The web plugin technology can be possible to directly execute the executable binary, written by C or C++, on the web browser. Therefore, it can protect source code in easy because the binary file is being used to distribute the application. However, even though the binary file in a web plugin is distributed though the web browser, it is executed on the real OS and the specific CPU. That means it is a platform-dependent technology. Furthermore, if it is infected by a malicious code and executed, the damage will affect on the whole system because a web plugin such as ActiveX can access to the system directly.

Google has proposed Native Client (Nacl) [3] that solves the defect of existing plugin technologies. This is the same as using an executable binary. However, it differs in that it restricts the function to the limited access of the system which is defined and allowed at compile time. In addition, PNacl (Portable Native Client), an extended version of Nacl, solves the processor (CPU) dependency by using the intermediate language of LLVM as an executable binary. However, the dependence of OS and web browser has not been solved.

2.2 Web-Based Emulators

The structure of a web-based emulator is shown in Fig. 2. A general emulator can emulate the executable binary file built for the specific processor at the different platform. For example, an application built for an ARM processor can be executed on PC platform with Intel x86 CPU using the ARM emulator. Therefore, the platform dependence problem occurred in the web plug-in technology can be solved by applying the emulation technology to the web application. Recently the various emulators have been introduced and developed so that web applications on the web browser can be executed on the emulation environments of the diverse processors such as x86, ARM, OpenRISC, and RISC-V etc.

A web application executed in the web-based emulator is the same with the common execution binary file. It is an advantage that a general application can be executed immediately on the web browser without any web plugin. Only the terminal GUI allows access to the OS running inside the emulator. In other words, the web-based emulator, similar to an independent virtual machine, does not provide any function that can interface with the outside of the emulator. Therefore, it is impossible to directly execute a native code in the web application using the web-based emulator.

In the examination of various web-based emulators, jor1k emulator [4] has the advantages in the point of the speed and the development convenience. Thus, it can be a good candidate to be used as the execution engine. jor1k is the emulator of

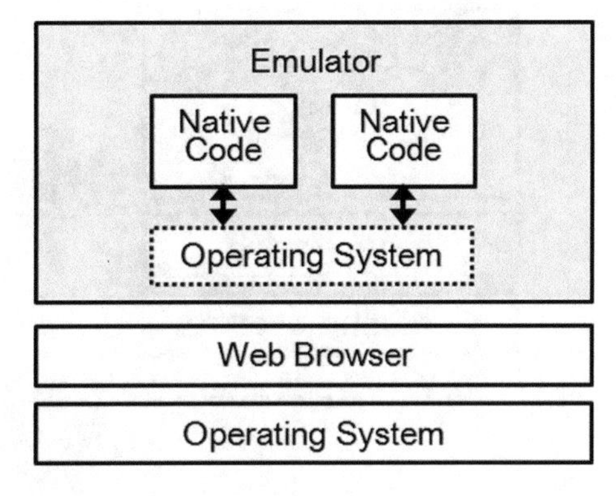

Fig. 2 The structure of a web-based emulator

OpenRISC OR1K CPU that is a processor core of Open Source. OS Kernel supports Linux version 4.4.0. jor1k provides the various utilities such as a compiler, a terminal and an editor etc. And the core of OpenRISC processor is relatively small, so it runs faster than other emulators. The jor1k improves the emulation speed by implementing the optimized asm.js code based on the Emscripten compiler [5].

3 Emulator-Based Web Application Execution Engine

In this paper, an execution structure of emulator-based web application is proposed in order to solve the source code protection problem that occurs in the script-based web application.

3.1 Emulator-Based Web Application Execution Engine

Web applications have been implemented script-based such as javascript or HTML, and Fig. 3a shows the structure of the common web application. Generally, there is a problem that the source code is exposed as it is when the web application is executed. Thus, the web plugin technology is used to prevent the leak of the source code and implement diverse functions in the web application. However, it is weak to the malicious code attack because a web plugin application code has been executed in OS layer directly. Also, it is dependent on the OS platform.

The Web-based emulator can execute the binary Native code independently on the web browser. Figure 3b shows the structure of Web-based emulator. However, the web-based emulator, like a virtual machine, is executed solely through the terminal of the interactive mode. Currently, the web-based emulator itself isn't cooperated with other web applications outside.

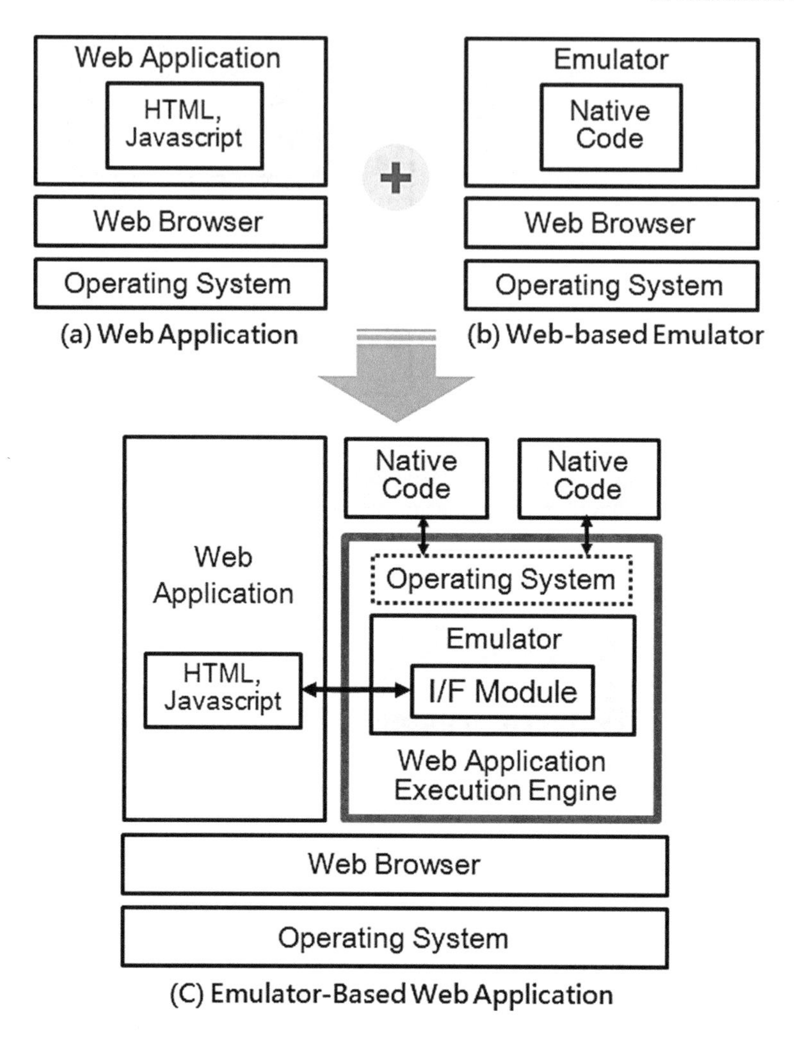

Fig. 3 The structure of an emulator-based web application

In this paper, the emulator-based web application execution engine is proposed in order to overcome the weakness of the source code protection in web applications, to easily implement the various functions of web applications, and to execute it fast. The proposed system is implemented using the web application execution engine based on a web-based emulator, which is called as the emulator-based web application engine.

The native code of a web application is directly executed on the emulator-based web application execution engine. Through this, the native code used in a web application is distributed as a binary format such as ActiveX in order to protect the source codes, and to solve the dependency of OS platform that is a problem of web-plugin methods.

The existing web-based emulator is another web application which is executed solely, and it doesn't provide a function that can interface with other web applications outside it. To implement the proposed emulator-based web execution engine proposed in this paper, the interface function of API level should be developed in order to linkage between the web-emulator and web applications.

Using the developed API, It is possible to directly load and execute native codes in the web application without any web plug-in technology. Fig. 3c shows the structure of the proposed emulator-based web application. The proposed structure is composed with the internal interface module of emulator that connects web applications and the emulator-based web execution engine, the kernel OS, and native codes that are executed in the execution engine.

The web application and the emulator-based web execution engine in the proposed system are executed on the same layer on the web browser, since the web-based emulator is another web application implemented by a web standard programming language such as HTML, Javascript, and CSS. Some functions in a web application, which does not want to be leaked, is implemented as executable native codes in a binary format. And they are distributed in the binary format with a web application code, unlike the script-based web applications.

However, the native code can't be executed directly on the web browser. Therefore, it is delivered through the proposed interface module and then executed on the web-application execution engine. In the proposed system, a native code cannot directly access the operating system or hardware because it is executed on the web browser layer. Even though the native code includes malicious codes, it cannot influence the operating system or hardware, resulting in the overall system safe.

3.2 *Emulator-Based Web Application Execution Process*

To execute the proposed emulator-based web application, the emulator-based web application execution engine should be downloaded and initialized different from the traditional method of the web application. Figure 4 shows the execution process of the proposed emulator-based web application.

First of all, the image file of the emulator-based web application execution engine and the image file of OS are downloaded with the web page code of a web application. After the downloading is finished, the initialization of the execution engine and the booting of the OS are executed to prepare for the execution of native code. If the initialization and booting is finished, it is possible to execute a native code and the system enters the waiting state to receive the native function call of an executed web application. In the waiting state, in case that the native code is called by the executed web application, it is delivered to the emulator-based web execution engine through the interface module.

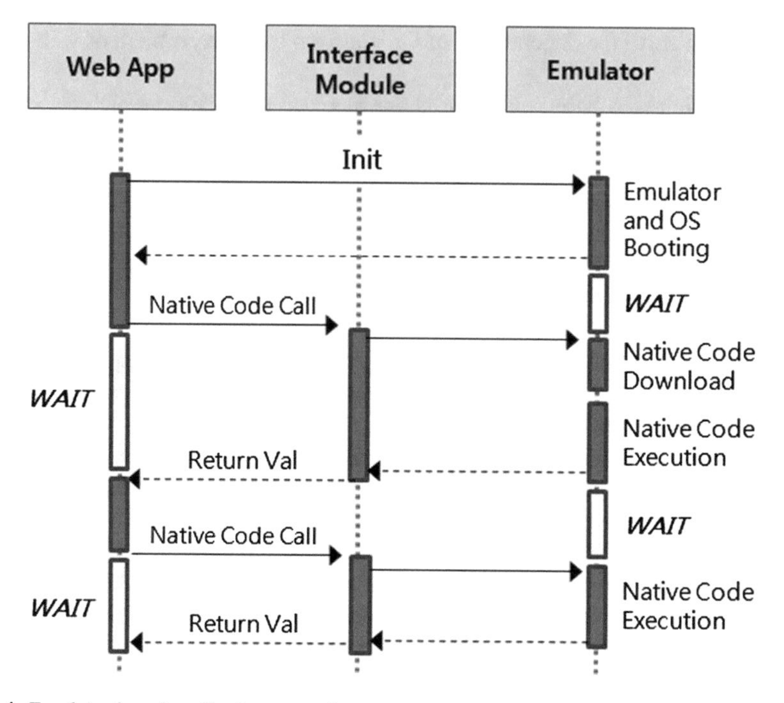

Fig. 4 Emulator-based application execution process

The emulator-based web application execution engine immediately download the native code from the web server and then load it. After that, the requested native code is executed on the emulator-based web application execution engine. If a native code is called several times, its download and load process are executed only once at first.

4 The Implementation of an Emulator-Based Web Application

4.1 The System of an Emulator-Based Web Application

Figure 5 shows the overall system of a web application applied the proposed emulator-based web application execution engine in this paper. In order to implement the system of an emulator-based web application, the interface module and the native code loader inside the web application execution engine and the interface API for a web application are necessary.

The interface module serves as a master which receives the call of the native-coded(binary) function from a web application and then requests the execution of it to the emulator-based web application execution engine. Also it receives the results

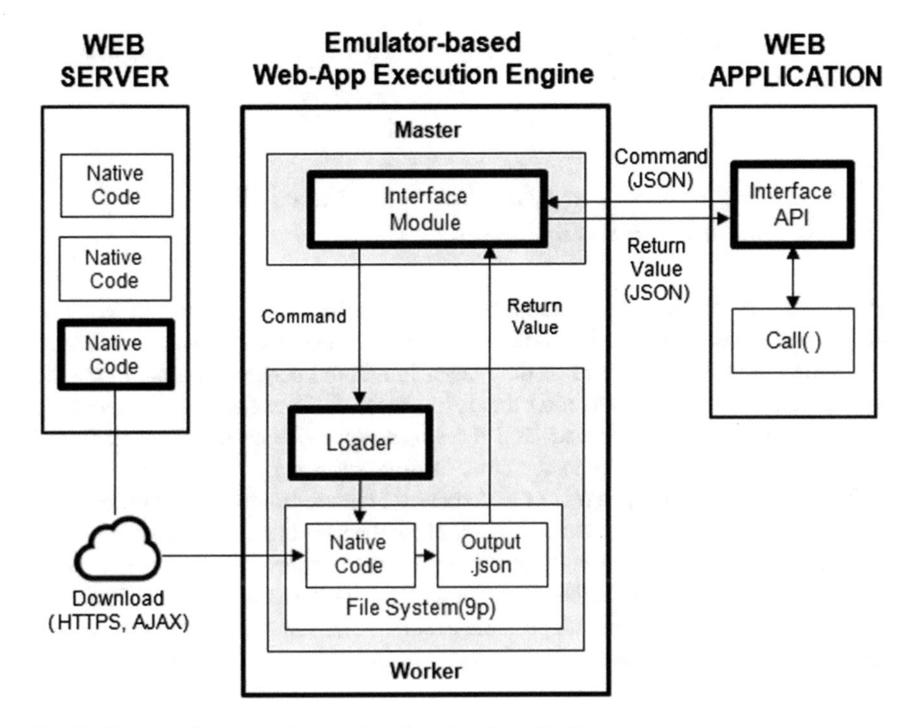

Fig. 5 The overall system of an emulator-based web applications

from the emulator-based web application execution engine and then return them to the web application. The interface module makes it possible for the emulator-based web execution engine to linkage to outsider web applications using the interface API.

The emulator-based web application execution engine serves as a worker which receives and executes the service request of a native code from the interface module and returns its results. The native code loader is in charge of managing native codes and executing them inside the emulator-based web application execution engine like a daemon program.

The native code can be a general executable file or a library file in .so format. If a native code is an executable file, it is executed on the internal terminal that is provided by the emulator. However, if the native code is in a library file, it is supported by the OS booted in the emulator. When the inner OS is a linux, it calls using the dynamic link library [6].

After the execution of the called native code is all completed, the result of the execution is saved in the output.json file. After the output.json file is created, the interface module calls the callback function. The relevant callback function checks the result through the returned json file. The json file is saved in the inner file system of the emulator, and it can be identified in the web application using the interface API. Also, the format of a json file isn't restricted as a specific one, and it can be developed in a freeform where developers of a native code and developers of a web application want to make it.

4.2 The Implementation of an Emulator-Based Web Application Execution Engine

Figure 6 shows an example of implementing the main inner components based on the jor1k emulator as the execution engine. The emulator-based web application execution engine is comprised of the master that is in charge of an interface with the web application and the worker that is in charge of the execution of native code on an emulator core. The master and the worker runs like an independent thread respectively, and communicate each other using a message.

The master thread plays a role of a wrapper for the user interface. In comparison, the worker is a thread that executes a core logic of the web-based emulator. Basically, the inner structure looks similar to the common emulator structure, but both are different in that it is executed on the web browser. The interface module is implemented in the master thread that makes it possible to communicate with the web application, and a loader is implemented in the worker thread which executes native codes.

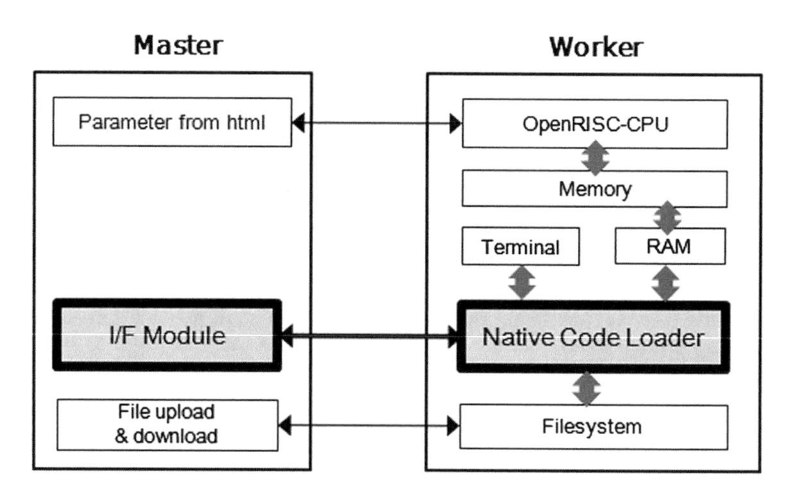

Fig. 6 The structure of an emulator-based web application execution engine using JOR1K emulators

4.3 Interface Module and API

The interface creating function, which is implemented as CreateInterfaceFunc() in the proposed interface API, generates an interface to call a native code in a web application. The argument values of the interface API are the information of the native code delivered, and the return of it is the result of the native code. When the interface API is called, the information of the argument value is written in the json format as shown in Table 1.

The type and filename among the argument information are expressed explicitly. The type field means a type of native code, and it can be classified into a common executable file case and a dynamic library file such as .so file.

The URL field means the internet address of a web server where the native code is. If the URL filed is identified, the native code is downloaded as a json file to a designated directory of the inner file system if the native code is not ready. Finally, the callback function is a handler function that processes a return value after the execution of native code is finished.

4.4 Native Code on an Emulator-Based Web Application

The native code is an executable binary file that is generated using a compiler. In general, executable binary files are only performed in the specific OS and CPU. On the other hand, the native code of the emulator-based web application execution engine can be executed on the various web browsers that follow the criteria of the web standard regardless of OS. One restriction is that the native code should be delivered to the emulator-based web application execution engine by the interface API of the web application.

A native code can be generated by a general compiler such as gcc. In the jor1k emulator, since OpenRISC compiler is supported internally, the source code can be directly compiled. In case of the developing environment like a PC, it can be compiled using the cross compiler.

Table 1 Arguments information of interface API

Name	Value type	Description
Type	String	The type of a native code – "EXE": executable binary – "LIBRARY": shared library
URL	String	The location of a native code on web server
Filename	String	Filename of native code
Callback function	String	Callback function called after executing a native code

Also, the code can be protected against the leak of source codes because it is distributes as a binary format. Furthermore, there is no dependency of the web browser or other OS because it is executed based on the emulator-based web application execution engine.

The native code has the following advantages on the emulator-based web application execution engine.

- The leak of source code scan be avoided because they are distributed as the binary files compared with the distribution of source codes in the script-based web application.
- The executable binary based on the source codes of C and C++ can be executed on the web application without any web plug-in. Therefore, the existing source codes of C and C++ can be used when the web application is developed.
- Different from ActiveX, even though a native binary infected by a malicious code is executed on the emulator-based web application execution engine, its damage is restricted inside the web browser, such as a sandbox. That is, although the web application is infected by malicious codes, OS and H/W are not affected.

5 The Experiment and Evaluation

In this chapter, the proposed the emulated-based web application execution engine is applied to a web application in order to examine the possibility of the source codes leakage and to test its execution speed through the interface module. For this examination, the encryption function was implemented by the native code and applied to the encryption web application. The system structure of the emulator-based encryption web application implemented for the test and evaluation is shown in Fig. 7. The algorithm [7] of AES written by C language on the SERVER is compiled to the executable binary of OpenRISC, as a native code for a web application.

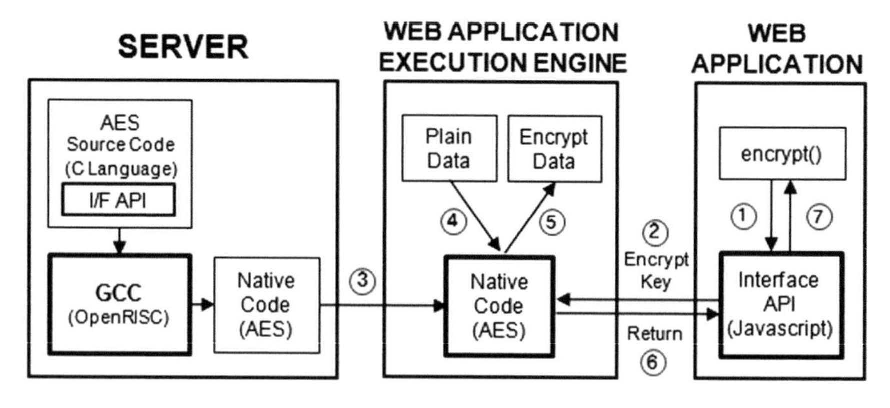

Fig. 7 The execution of an emulator-based encryption web application

5.1 The Performance Evaluation

The execution time of the experimental application is shown in Fig. 8. The execution time includes the time to call a native code by web application, the time to initialize a native code(download and load process), the time to boot the OS in the emulator-based web execution engine, and the time to execute it on the emulator-based web application execution engine. The native code used for this experiment, the algorithm of AES encryption, was executed to encrypt and decrypt the data of 64 KB with the ways of CBS and ECB a hundred times. The experiments are performed on 3 different platforms like a PC, iOS, and Android.

The experiment environment of each platform was as following: PC was a laptop with Intel i2540 m CPU, 8 Gb RAM, Windows 7, iOS was an Apple smart phone with iOS 11, and Android was a smartphone with Android 5.0.1 version. The detailed experimental environment is shown in Table 2. Also, Chrome is used as the web browser for all devices.

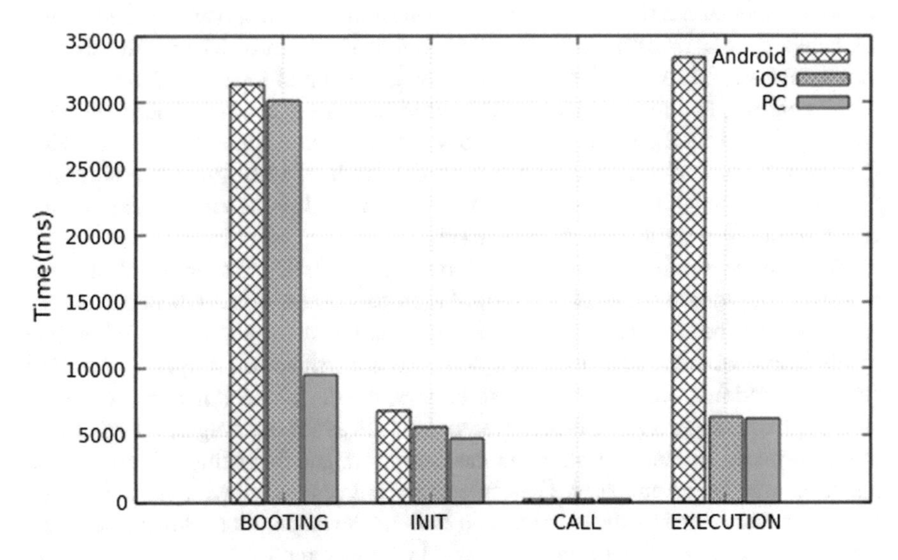

Fig. 8 Emulator-based web application performance

Table 2 Experimental environment

Type	PC	iOS	Android
OS version	Window7 Service pack 1	11.3	5.0.1 (Lollipop)
Processor	2.6 GHz (i5-2540 M)	1.4 GHz (Apple A8)	2.3 GHz (Snapdragon 800)
Memory	8 Gb	1 Gb	2 Gb

As shown in Fig. 8, the PC outperforms to others according to the booting time, since the hardware specification is critical to the performance. For the execution time of the native code, the iOS and the PC show competitive each other and superior to the Android, since the Chrome on the iOS is well optimized browser compared with it on others. However, there is no significant differences among 3 platforms in the comparison with time of the initialization and the call of a native code, since the I/O-bounded works such as the code download or network communication were less affected by the performance of system. In these experiments, a native code implemented with a source code was executed on 3 different platforms with 3 different OSs and CPUs. That confirmed that the proposed system in this paper was platform-independently executed.

5.2 *The Source Code Protection*

To check whether the source code downloaded from the web server is leaked or not, the native code is investigated with "file" command of Linux. The results showed that the native code was a format of the executable file of OpenRISC with 32-bit, which confirms that the native code was not leaked. Also, in other experiments, the native code included only the header information of the format of ELF execution file and the code of machine language as the result of identifying the files with "hexdump" command. Therefore, they verified that the leak of source codes did not occur, resulting in protecting the source safe.

Through the web browser developer tools, the possibility of source code leakage is tested. The source codes of the script-based part in the web application and the emulator could be read and identified. However, the native code couldn't be identified by the developer tools, since the native code was not written in a script language like Javascript and in a binary format. Furthermore, the downloaded native code is not stored like cookies or a general-purpose storage such as a web storage. Instead of that, in the proposed system, it is stored in the file system and the memory implemented inside the web-based emulator. Therefore, the resources related with the native code couldn't be identified by the existing web browser developer tools. After closing the web application, the native code disappeared with the emulator.

6 Conclusions

In this paper, the emulator-based web application execution engine was proposed. Using the proposed one, the web source code leakage of the web application was prevented by the execution of the native binary code. Also, the emulator-based web application was executed independent of platforms different from the web plug-in technology. We implemented the execution engine-based application using the interface API.

Experiments showed the proposed approach could be applied to the practical web application. In experiments, we verified the source code protection ability of the proposed approach. The performance of the proposed one showed the execution time of the native code was fast enough. However, other overheads like the booting time and the initialization time of a native code increased the overall execution time.

For the further work, we will decrease the booting time to improve the execution speed, and upgrade the interface module so that it can be used for the various projects. Especially, we will verify the possibility of expanding to the various fields such as the client web storage for security [8] as the isolated space. Finally, we will continue the research about the plan for the modularization and the improvement of performance that can be used as the general-purpose.

References

1. Malzilla. Available: http:// http://malzilla.sourceforge.net/
2. K-Apps: Method for providing encrypted web application, terminal supporting the same, and recording medium thereof. KR Patent 1014723460000, 2014
3. Native Client. Available: https://developer.chrome.com/native-client
4. Jor1k. Available: http://s-macke.github.io/jor1k/
5. Zakai, A.: Emscripten: an LLVM-to-JavaScript compiler. In: Proceedings of the ACM International Conference Companion on Object Oriented Programming Systems Languages and Applications Companion, pp. 301–312
6. Dynamic loading library. Available: https://en.wikipedia.org/wiki/Dynamic_loading
7. Tiny-AES. Available: https://github.com/kokke/tiny-AES-c/
8. Kurumatani, S., Kasae, Y., Toyama, M., Akama H.: Hosting a server on a browser using Wemu architecture. In: Proceedings of the 16th International Conference on Information Integration and Web-based Applications & Services, pp. 159–162 (2014)

A Study on the Influence and Marketing Effect of Korean Wave Events and Festivals Organization

Jae Ho Park, Jeong Bae Park and Cheong Ghil Kim

Abstract In recent years the global popularity of Korean Wave is increasing in entertainment areas and spreading to other regions of the world. As a result, there have been many organizations formed to support its creative industries; at the same time, the field of the satisfaction and perceived effectiveness of them has emerged as an important issue. This paper introduces a method of measuring the influence and marketing effect of organizations for Korean Wave Events and Festivals. For this purpose a classical test theory of Cronbach's alpha is utilized. The feasibility of measuring results is also ensured by exploratory factor analysis. The result shows 0.7 or higher alpha values and could provide suggestions to establish a theory of events and festivals for organizations by active follow-up studies for resolving will-less different results in this field due to a lack of the proceeding research.

Keywords Korean wave · Test theory · Cronbach's alpha · Statics · Festivals

1 Introduction

The popularity of Korean Wave has increased greatly in the fields of entertainment and culture since 1990s and recently this cultural trend is spreading to other regions of the world. This is a cultural phenomenon known as Hallyu. Especially, it came into vogue in Southeast Asia and mainland China. For example, it is very popular among young people enchanted with Korean music (K-pop), dramas (K-drama), movies,

J. H. Park · J. B. Park
Department of Performance Planning and Management, ChungWoon University, 32244 Hongseong, Choongnam, Korea
e-mail: event987@naver.com

J. B. Park
e-mail: pjb@chungwoon.ac.kr

C. G. Kim (✉)
Department of Computer Science, Namseoul University, 31020 Cheonan, Choongnam, Korea
e-mail: cgkim@nsu.ac.kr

R. Lee (ed.), *Software Engineering Research, Management and Applications*, Studies in Computational Intelligence 789, https://doi.org/10.1007/978-3-319-98881-8_5

fashion, food, and beauty in China, Taiwan, Hong Kong, and Vietnam, etc. This has been closely connected with multi-layered transnational movements of people, information and capital flows in East Asia [1–3].

Surprisingly, within just a decade, Korean dramas have become one of the types of broadcasting content most in demand in many Asian countries and this movement has broaden to other reasons [3, 4]. According to the World Trade Organization (WTO), Korea became the second major exporter among Asian countries, following Hong Kong. As for the economy size, the Korea Creative Content Agency (KOCCA) reported that the export amount reached $6.85 billion with the growth on average by 8.5% more than last year. Figure 1 shows the percentage of content export by country [5]. Exports of Korean cultural content to Asia have steadily increased since 2012; they became more than half of the total in 2014, accounting for 57.4%. The export portion of the North America and Europe is close to 20% even though those of Asian countries are still high.

Under these circumstances, there have been many organizations formed to support this creative industries locally. Especially, festive utilization policies have been established to revitalize regional events by local governments. According to the statics of Korean government the total number of festivals in Korea is 758, and the 32.7% of them are related with local culture and art. And the government budget to support the festival exceeded 220 billion Korean Won [6]. At the same time, the field of the satisfaction and perceived effectiveness of them has emerged as an important issue. In general, local festivals are hosted by local governments and various local groups for various purposes such as strengthening the image of the region, creating regional economic value, and providing economic value to the local people. Furthermore, they also provide the local core contents to the tourist through the unification of the local residents.

Domestic regional festivals are held in various types and methods throughout the country. According to the Ministry of Culture, Sports and Tourism, more than 1300 festivals have been held. These festivals are organized and operated by professionals, volunteers, or by corporations and non-profit organizations. However, in general, the organization of local festivals are complex and combination of special elements are linked together, which is why professional theories and methods are needed. Therefore, the festival organization should learn to think and act systematically [7].

The system of festival organizations can be categorized as three groups. The first group is the administration-led organizational system that implements various admin-

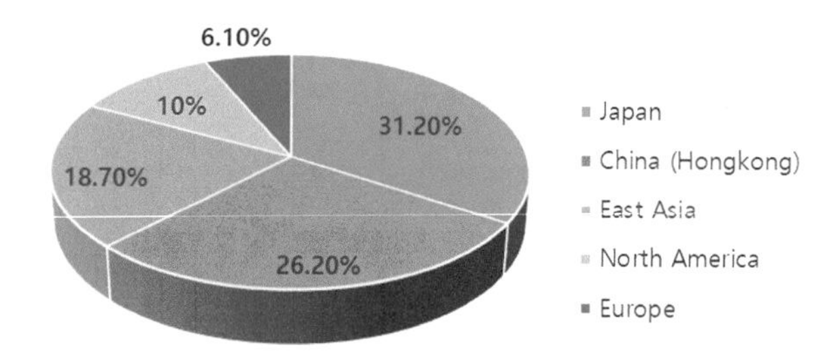

Fig. 1 Percentage of content export by country

istrative measures; the second is sponsored by a company and led by local residents; the third type is administration and private sector led, where citizens and administration participate in various forms. Such organizational system must be interdependent and the interacting elements should be unified into a system. Therefore, festival organizers need to continue doing research and development in order to achieve a purpose of creating economic and social benefits and values through strategies that can allow to understand the dynamics of interdependence and to predict future coming changes.

In case of professional festival organizations, the roles are divided as follows: chairman or secretary for administration part and general director for composition and direction of program. And for running stage, a department shall be established to manage administrative affairs, public relations, facilities, directing, operation, etc. At this time, it is necessary for each side to clarify and establish clear lines of authority and responsibility in order to maintain the good cooperation system which may lead to a successful festival. Recently, festival-related organizational structure has been systemized to some extent and the development and operation of festivals are being carried out in order to establish effective performance by recognizing the concepts and methods associated with a systematic approach. However, there might be a problem of suffering a shortage of skilled professionals and trained volunteers in a small number of non-profit organizations. For this reason, a need for research on how the actual festivals are perceived when it comes to effectiveness and satisfaction of general management and marketing plan aspects.

In this work, we introduce a method of measuring the influence and marketing effect of organizations for Korean Wave Events and Festivals. For this purpose a classical test theory of Cronbach's alpha [8] is utilized. The feasibility of measuring results is also ensured by exploratory factor analysis. The rest of this paper is organized as follows. Section 2 summarizes the theoretical consideration of the study. Section 3 introduces our research design and analysis method. The conclusion is covered in Sect. 4.

2 Use Theoretical Consideration of the Study

2.1 Preliminary Study of Festival Organization

In general, the issue who will be the organizing party of a festival is important anything else. It could be either a profit or non-profit organization. According to this classification, a festival will be hosted by either 'government (admin)' or 'people (private)' led. A festival organization can be composed of those who are responsible for various aspect of a festival such as hosting, managing, planning, directing, and sponsoring.

This study suggests that it would be easy to revitalize the government initiated system considering budget, operation, and the responsibilities of the administration of public officials compared to limitation of private initiative; but at the same time, it argues that such systems can undermine diversity, autonomy, and creativity introduced by a report of Korea Cultural Policy Development Institute in 1994. It is reported that there are many festival openings by public-private cooperation. A government report claims that the form of public-private partnership is dominant stating that 38 out of 44 cultural tourism festivals in 2010 were jointly led by public-private form, which is 86% of whole.

However, in terms of budget and administrative roles, it seems that public administrations are playing leading roles, but the important thing is that both parties take advantage of other's roles to achieve a successful outcome. The success factors of a festival could be summarized in terms of efficiency, voluntary participation with unity, and self-esteem through improvement of the quality of life and culture of local residents. For this purpose, a festival organization must make efforts to promote the effectiveness of an event. The fact that festivals constitute a committee and systematic organization today is an efficient operation strategy to achieve a certain goal. Festival organizations usually have the functions such as strategic decision-making, goal management, budget planning, and the secretariat. They should be professional, which is responsible for the overall execution of the festival [9].

The festival organization plays a major role for providing appropriate functions and positions to make a successful festival. At this time, satisfaction and perceived effects become very important. As for local festivals, they are held for image formation, residents' harmony, local economic ripple effect, and more. For traditional local festivals with future oriented vison, the festival organizational structure and its satisfaction and perceived effect are playing important factors for making the right scale and characteristics of the festival in such a way that the impact will be the greatest, the organization will be thorough, and the participants will be satisfied.

2.2 *Satisfaction, Previous Studies of Perceived Effects*

It is known that satisfaction can be a key factor in understanding the behavior of a person when the role recognition for the partner matches the expected role in the relationship of sponsorship, which provides an important motivation for driving the identity of an organization [10, 11]. Leading variables affecting satisfaction could be expectation, performance, and comparison of expectation versus performance, emotional response, and fairness [12]. Also, it may influence sponsorship behavior indirectly according to a nonprofit relationship marketing research [13].

In outsourcing relationship, perceived effect is the result of communication that produces mutual sharing performance while continuing relationship by reducing conflict with mutual trust and commitment. In particular, relationships can be maintained by mutual trust and commitment in relationship exchanges. At this moment, the performance improvement of a business relationship can take place [14]. This represents the customer's perspective on the acquired effect in an outsourcing relationship. The perceived effect influences the whole relationship development process of recognition, exploration, expansion, and binding [15]. Therefore, it is possible to achieve mutual sharing of information through frequent exchange of information between two parties to increase the sustainability of the relationship and to reduce the conflict of functional disorder. The definitions of satisfaction and perceived effects are summarized as shown in Table 1.

3 Research Design and Analysis Method

To carry out this research, we have modified the concept of festival organization, satisfaction, and perceived effect including some items based on previous research data and some hypotheses have been set about the satisfaction and perceived effect of the festival organization.

Table 1 Definition of satisfaction by researcher

Division	Researcher	Definition
Satisfaction	Covin et al. [11], Mael and Ashforth (1992)	Satisfaction is an important factor driving the identity of a company or organization
	Fornell et al. (1996)	Many agencies focus on satisfaction to keep consumers on track, and use it as a measurement tool for performance
	Szymanski and Henard [12]	Leading variables that affect satisfaction are expectation, performance, comparison of expectation versus performance, emotional response and fairness
	Szymanski and Henard [12]	A factor influencing indirect sponsorship behavior in relation marketing of nonprofit organizations
Perceived effect	Dwyer et al. [15]	Relationship development stage: recognition, search, extension, and bond process
	Grover et al. (1993)	The perceived effect represents the customer's perspective on the effects of outsourcing relationships
	Morgan and Hunt [14]	The relationship can be maintained by mutual trust and commitment in the relationship exchange. At this time, increase the performance of business relations
	Mohr et al. (1996)	It is possible to achieve mutual sharing of information through frequent exchanges of information between two parties, to increase the sustainability of the relationship, and to reduce the conflict of functional disorder

The subjects (targets) of this study were 400 workers being directly or indirectly engaged in festival related companies, and they were surveyed about the satisfaction and perceived effect of the festival organization. A statistical analysis of the questionnaire was conducted by analyzing the general characteristics and the average and frequency. To secure reliable calculation and validation of Cronbach's alpha coefficients, the reliability and validity of the instrument were verified by exploratory factor analysis. In addition to verify a single dimension and internal consistency of the variables included in this study model, the verification was carried out in order to increase the validity and reliability of each variable measured by the subjective cognitive scale, and to achieve specific research objectives. The Cronbach's alpha coefficient has a value between 0 and 1 and the relationship between levels and reliability are introduced in Table 2. In case of exploratory analysis, if it is 0.6 or more, it is evaluated that the reliability is satisfactory level. In general, the reliability is good when it is 0.7 or more, and it is excellent when it is 0.8–0.9 or more. In addition, multiple regression analysis was conducted to verify the hypothesis of this study model and the hypothesis relation was verified by using the analysis method of the research model.

4 Empirical Analysis

4.1 Reliability and Validation of Measurement Tools

Table 3 shows the results of the basic statistics and the Cronbach's alpha coefficient for the satisfaction of the festival organization and perceived effect factors. Among the items, 'festival organization increases the sustainability of the relationship' was 3.92 points, which showed the highest score. 'Festival organization is satisfied with conflict resolution' scored 3.65, which was relatively low compared to other items. The overall reliability was 0.930, which showed excellent reliability, and there were no items that would impair reliability.

Concentration validity is about the degree to which two or more measurement tools correlate to a constituent concept. In this paper, we confirm the magnitude of standardization factors, the average variance extraction value, and the reliability of the constituent concept to verify the convergence validity. For the variables that measure constructs, it is claimed that when factor loadings, i.e. parameter estimates are high and statistically significant, there is intensive validation between constructs.

Table 2 Cronbach's alpha coefficient (0–1)

Level	Reliability
0.6 or more	Satisfactory
0.7 or more	Good
0.8–0.9	Excellent

Table 3 Satisfaction, reliability analysis of perceived effects

Item	Item analysis		Item—Overall correlation Factor	Reliability	
	Average	Standard deviation		When removing items Cronbach's α	Overall reliability
Enables information sharing and exchange of festival organizations	3.78	0.79	0.641	0.929	0.930
Reduce the risk of festival organization	3.80	0.95	0.645	0.929	
Increase sustainability of festival organization relationships	3.92	0.82	0.693	0.927	
Satisfied with organizational commitment of festival organization	3.88	0.80	0.761	0.924	
Satisfied with organizational trust of festival organization	3.85	0.83	0.785	0.923	
Satisfies the risk reduction of the festival organization	3.77	0.83	0.755	0.924	
Satisfies conflict solving through festival organization	3.65	0.90	0.770	0.923	
satisfied with communication enhancement through the festival organization	3.81	0.84	0.777	0.923	
Satisfied with staff and team development through festival organization	3.79	0.82	0.749	0.925	

Therefore, in the confirmatory factor analysis, standardized parameter estimate 0.6 or more is desirable. As shown in the Table 4, the parameter estimates for festival organization, satisfaction, and all measured variables of the perceived effect were 0.7 or higher, which has a significant statistics. Therefore, it can be said that there is intensive validity of measurement tools for festival organization, satisfaction, and perceived effects. Also, since AVE (average variance extracted) means the average of squared values of the standardized parameter estimates. If the value is 0.6 or more, it is considered to have an intensive validity. Finally, the construct reliability value can be calculated as the standardized parameter estimate and the error variance. If the value is 0.7 or more, it is considered to have intensive validity. In this paper, it can be seen that the convergence validity is secured since the reliability of all constructs concepts is 0.7 or more.

4.2 Hypothesis Verification

Since the significant regression coefficients are all positive when looking at the relationship between satisfaction and perceived effectiveness, it can be seen that the organization commitment and job satisfaction are high when both satisfaction and perceived effect of festival organization are at a high level shown in Table 5.

According to the simulation the effect of the festival organization on the satisfaction had a significant effect on the satisfaction level of the festival organization under the significance level of 1%. Looking at the standardized path coefficients, they are marked with (+) signs, which may increase the satisfaction of the festival organization. Also, when comparing the values of the standardized path coefficients, the result shows that it has greater influence on a festival organization's satisfaction level. As a result of analyzing whether there is a significant effect on the perceived effect of festival organization, it can be seen that there is an influence at a significance level of 1–5%. As the standardized path coefficient shows the influence relationship of (+), perceived effect is expected to be enhanced. Also, when comparing the values of the standardized path coefficients, the figure shows greater influence on the perceived effect. The hypothesis test results show that the hypothesis is adopted including the hypothesis H1, H2, H3, and H4. The result of the research is predicted as the concept of festival organization, satisfaction and perceived effect were already recognized by the author. However, it is necessary to establish a theoretical foundation of the festival organization because the researches in related fields are lacking. It also suggests that continuous research is needed through business and exchange of related fields in the future.

Table 4 Satisfaction, reliability analysis of perceived effects

Factor	Measurement variable	Standardized parameter estimates	t-value	Distributed extraction volume	Configuration concept reliability
Satisfaction	Satisfied with organizational commitment of festival organization	0.739	–[a]	0.673	0.710
	Satisfied with organizational trust of festival organization	0.799	10.01$^{###}$		
	Satisfied with risk reduction of the festival organization	0.854	10.72$^{###}$		
	Satisfied with conflict solving through festival organization	0.860	10.78$^{###}$		
	Satisfied with communication enhancement through the festival organization	0.861	10.80$^{###}$		
	Satisfied with staff and team development through festival organization	0.802	10.05$^{###}$		
Perceived effect	Enables festival organization's information sharing and exchange	0.798	–[a]	0.665	0.708
	Reduces risk factor of the festival organization	836	10.48$^{###}$		
	Increases sustainability of festival organization relationship	0.812	10.18$^{###}$		

$^{###}p<0.01$

[a]Can not be calculated because we have set 1.0 to fix the configuration concept variance

Table 5 Relationship between festival organization and satisfaction, perceived effect

Route	Standardized path coefficient	t-value
Organizational Commitment of Festival Organizations → satisfaction level	0.413	$9.29^{\#\#\#}$
Job satisfaction of festival organization → satisfaction level	0.407	$9.29^{\#\#\#}$
Organizational Commitment of Festival Organization → perceived effect	0.411	$8.87^{\#\#\#}$
Job satisfaction of festival organization → perceived effect	0.379	$8.19^{\#\#\#}$

$^{\#\#\#}p<0.01$

5 Conclusions and Implications

This study introduces the effect of festival organization satisfaction and perceived with the aim of suggesting the important factors for creating performance through human capital and organizational management in the festival organization. In order to accomplish the purpose of this study, we tried to verify the theoretical and empirical effects of satisfaction and perceived effects on the related workers of the festival organization.

For this, literature review and empirical research were concurrently conducted. Through the literature review, the relationship between festival organization, satisfaction, and perceived effects was clarified and systematized.

To verify the reliability of measurement tools for satisfaction of festival organization and perceived effect, Cronbach's alpha coefficient was calculated, and the reliability was good as it showed over 0.7. In addition to this, exploratory factor analysis was conducted to verify the validity of the measurement tools and the validity of the results was also confirmed.

Four hypotheses were constructed and analyzed for the satisfaction of the festival organization and the significant influence on the perceived effect. The higher the organizational commitment and the job satisfaction are, more influence satisfaction and the perceived effect of the festival organization had on the degree of mutual relationship.

It needs to be pointed out that there is a need for continuous research in the academic perfection since the field research is lacking compared to the interest of the business. Therefore, despite the implications of this study, the following limitations exist. First, for verification, more specific and systematic study on the concept and satisfaction, and perceived effect of the festival organization is needed. Festivals have a continuity that is held annually at a certain time so for festivals to be culturally and socially engaged, a management organization is needed in order to activate cultural values and communication structures throughout the year.

Therefore, it is meaningful to study the applicability of festival organization in terms of satisfaction and perceived effect considering the recent festival performance measurement. However, additional research and approaches seem necessary. In addition, we attempted to derive empirical factors for the interaction of festival organization satisfaction and perceived effects in this study, but it has limitations that can't be verified for the variables that can occur in actual festival environment. To overcome limitations of this study, it is wished that future studies will continue on various theories and achievements about the festival organization.

References

1. Shim, D.: Asian popular culture and Korean wave. Asian Commun. Res. **12**(2), 5–9 (2016)
2. Jeon, W. K.: The Korean wave and television drama exports, 1995–2005. Ph.D. thesis, University of Glagow (2013)
3. Kwon, E. J.: Korean Wave: Discourse analysis on Korean popular culture in US and UK Digital Newpapers, M.A. thesis, Radbound University (2017)
4. Kim, B.: Past, present and future of Hallyu (Korean Wave). Am. Int. J. Contemp. Res. **5**(5), 154–160 (2015)
5. Global Hallyu Issue 2017, Monthly Hallyu Report, Vol 129, KOFICE (2017)
6. Lee, H.: Utilize festivals to revitalize local tourism. Hanyang University at http://kto.visitkorea.or.kr/kor/notice/data/report/org/board/view.kto?id=418419&rnum=5 (2013)
7. Getz, D., Frisby, W.: Evaluating management effectiveness in community-run festivals. J. Travel Res. **27**(1), 22–27 (1988)
8. Cronbach, L.J.: Coefficient alpha and the internal structure of tests. Psychometrika **16**, 297–334 (1951)
9. Popescu, R.I., Corbos, R.A.: The role of festivals and cultural events in the strategic development of cities. Recommendations for urban areas in Romania. Informatica Economică **16**(4), 19–28 (2012)
10. Ashforth, B.E., Mael, F.: Social identity theory and the organization. Acad. Manag. Rev. **14**, 20–39 (1989)
11. Covin, T.J., Sightler, K.W., Kolenko, T.A., Tudor, R.K.: An investigation of post-acquisition satisfaction with the merger. J. Appl. Behav. Sci. **32**(2), 125–146 (1996)
12. Szymanski, D.M., Henard, D.H.: Customer satisfaction: a meta-analysis of the empirical evidence. J. Acad. Mark. Sci. **29**, 16–35 (2001)
13. Arnett, D.B., German, S.D., Hunt, S.D.: The identity salience model of relationship marketing success: the case of nonprofit marketing. J. Mark. **67**(2), 89–105 (2003)
14. Morgan, R.M., Hunt, S.D.: The commitment-trust theory of relationship marketing. J. Mark. **58**(3), 20–38 (1994)
15. Dwyer, F. R., Schurr, P. H., OH, S.: Developing buyer-seller relationships. J. Mark. **51**(2), 11–27 (1987)

Understanding the Success Factors of R&D Organization

Donghyuk Jo and Jongwoo Park

Abstract The importance of Research & Development (R&D) capability is increasingly more emphasized in recent year as corporations try to enhance its competitiveness to changing market in ever more fierce competition. R&D is core activities that sustain organizational innovation and the R&D team is regarded the key component of R&D organization. In today's rapidly changing business environment, the success of the organization depends critically on whether it has the core capital needed for its business success and how it can use it effectively. Therefore, the purpose of this study is to understand the contributing factors of R&D project in terms of social capital perspective, which is being considered key resource of business management today. Social capital, knowledge sharing, and Team-Efficacy are presented as core element of R&D team performance based on the review of previous literatures and empirical test of verified relationships. The significance of this study is in validating the importance of team capital and competence building under R&D project environment and presents strategic direction for R&D project success and team competence.

Keywords R&D project · Team capital · Knowledge sharing · Team-efficacy R&D performance

1 Introduction

Social capital is the aggregate of real or potential resources embodied in a network of relationships of an individual or social structure [1], where the individual or social structure that constitutes the network accesses and utilizes resources inherent in the network. And increases their values [2]. Corporations today are forced into virtually endless competition due to accelerated pace of technology, shortened product cycle, and more fierce global competitiveness. The ability to develop creative ideas and new products by responding to the ever-changing market demand is viewed as crucial to

D. Jo (✉) · J. Park
Department of Business Administration, Soongsil University, Seoul, South Korea
e-mail: joe@ssu.ac.kr

R. Lee (ed.), *Software Engineering Research, Management and Applications*, Studies in Computational Intelligence 789, https://doi.org/10.1007/978-3-319-98881-8_6

business success [3–5]. Enormous investment is being made by corporations today to reinforce R&D competence and secure sustained competitive edge. However, there is a limit to R&D resource to be invested, which makes systematic and strategic R&D management ever more importance for reinforcement of corporate competitiveness [4, 6]. Innovative products are an importance source of competitive edge and corporations must sustain innovative efforts to remain competitive [6, 7]. R&D refers to all creative activities performed to create new knowledge and use them systematically to produce new products and services [8]. R&D is the core activity needed to maintain innovative organization [7, 9–13].

R&D projects are knowledge-intensive and all team members must have expertise in order to develop new technologies or products. If individual team members can't share their individual knowledge appropriately, the team as a whole cannot use the knowledge in full capacity. The best solution to the problem can be obtained when everyone has the common knowledge gained from knowledge sharing process. Therefore, R&D capacity is the ability of the team to share their knowledge in the process of performing R&D project to create, develp and realize creative solution [7, 10, 11].

Traditionally, the preceding studies considered the physical capital and human capital as the key factors that will determine the organizational performance. In recent years, however, the social capital attract our attention in that the more the social capital such as the relationship, trust and empathy between organizational members, the more likely that the companies achieve successful results [4, 9, 14, 15].

Social capital is seen as a key determinant of team competence in creating innovative performance [4, 16]. The social capital increases trust between organizational members, which, in turn, causes them to share knowledge and accumulate a wide range of experiences, and thereby they will have confidence in their organization, resulting in a positive effect on the organizational performance ultimately [17, 18]. Through the formation of a common consensus of the social capital, the members establish strong trusting relationships with the working members and cooperate with each other to achieve the goals of the organization. Therefore, in the R&D project environment, team members must work together to share insights and knowledge and transform innovative ideas into viable processes, products or services [15].

When team members perceive that their team has sufficient ability to accomplish the given tasks, they will develop a higher level of emotional attachment to the team. Also, the team can use such sense of loyalty to make various efforts to enhance performance [19]. In other words, team-efficacy plays a significant role in team members supporting and cooperating with each other and in forming positive and active interpersonal relationships, and helps improve team performance [20].

An organization should be flexible and responsive in order to provide successful products in a rapidly changing business environment. Therefore, having the core capital needed to achieve business objective and how such capital can be used effectively is what determines the performance of organization performance in today's rapidly changing environment [4, 7, 21].

The purpose of this study is to understand the contributing factors to R&D project success from the point of view of social capital, which is emerging as the core resource of today's business management. Review of previous literatures was done to analyze the relationship among different variables such as social capital, knowledge sharing and team-efficacy that determine R&D team performance. Finally, the study will recommend strategic suggestions to help R&D project succeed and enhance team competitiveness.

2 Theoretical Background and Hypotheses

2.1 Team Social Capital

Social capital is the aggregate of real or potential resources embodied in a network of relationships of an individual or social structure [1], where the individual or social structure that constitutes the network accesses and utilizes resources inherent in the network. And increases their values [2].

Social capital consists of three sub-dimensions: structural dimension, relational dimension, and cognitive dimension [1]. The structural dimension is the pattern of connection that forms a network and focuses on the type of connection, structure (density, centrality, class) and the multipurposeness. The structural dimension is social capital that appears in the nature or form of the bond between members. The members of the organization can access knowledge and resources through social network [18]. The cognitive dimension refers to the cognitive system shared by the members of the social network. The cognitive dimension is a shared cognitive system in which the members of organization shares common vision, values, goals, language, or semantic system to understand common meanings among network members. In other words, the cognitive system in the organization is the basic mechanism that maintains the consistency of behavior within the organization and preserves and develops the norms, values, and culture of the organization [1]. The relationship dimension is the quality of interpersonal relationships formed by the individuals who form a social network and is explained by trust, duty, or expectation that ensures social interaction [1, 2, 22]. Trust in members of the social network contributes to organizational performance by promoting cooperative activities and reducing conflict and transaction costs [23].

The structural capital (network) in the structural relationship between sub-dimensions of social capital affects the formation of cognitive capital (shared vision). The social network is interconnected through social relationships among the members of society [1, 4]. Castro and Roldán [24] stated that personal cognitive capital is created through frequent interactions that share the same practices that enable learning of skills, knowledge, and common behaviors. In this paper, we propose a new methodology of social networking in Korea. In this paper, we propose a new methodology for social networking [1]. Structural capital (network) also influences the formation of relational capital (trust). In other words, trust relationships evolve, and it evolve through social interaction [25], and the frequency of contact affects emotional trust [26]. As such frequent contacts and interactions among network members play the role of promoting trust formation [27], structural capital can be expected to have a positive impact on relational capital, and cognitive capital (shared vision) affects the formation of relational capital (trust). In other words, sharing vision, value and code among network members facilitates the formation of trust relationships [1]. While members work for a common goal, they do not expect others members seek personal gain [22]. Since the trust of organizational members is based on the consistency of values, sharing value among members plays a role in promoting trust [27].

In this way, the formation of a social network has an important influence on the formation of common values and goals among the members of the organization, and the organization maintains various forms of networks, thereby allowing the members to share the vision and goals of the organization. It encourages norms, values and customs to form trust relationships [4, 22, 27, 28]. Therefore, this study established hypotheses as follows:

Hypothesis 1 Network will have a positive effect on Shared Vision.
Hypothesis 2 Network will have a positive effect on Trust.
Hypothesis 3 Shared Vision will have a positive effect on Trust.

In addition to being able to share knowledge about work through social capital, members of an organization can create or utilize new knowledge [1]. Members of the social network can access and utilize knowledge through the social network, and obtain information and resources [18, 22, 23, 29–31]. Moreover, network members can share their vision and value to enhance their cohesion with a common purpose and also have the same perception to reduce conflicts in communication and freely exchange ideas, knowledge, and resources [22]. In other words, when members share a common vision and value, mutually meaningful communication is promoted, knowledge of exchange and combination of knowledge is recognized, and sharing of knowledge among members is facilitated [18, 29–32]. And trust among network members is a factor that promotes joint effort and cooperative behavior [25]. which not only enhances access to knowledge but also promotes knowledge sharing by increasing expectation and motivation for the value of knowledge [1]. In other words, high trust among network members helps members to participate actively in social exchange process, thereby promoting exchange of knowledge, reducing knowledge exchange costs, and helping shared knowledge to be utilized [2, 18, 29–31]. Therefore, this study established hypotheses as follows:

Hypothesis 4 Shared Vision will have a positive effect on Knowledge Sharing.
Hypothesis 5 Trust will have a positive effect on Knowledge Sharing.

Team efficacy is a group-level belief that a team can perform a given task successfully and is a measure of the confidence in the group's ability and capacity [33]. Members of the organization form consensus and build trust relationship among themselves through social capital [34]. It is notable that the higher the trust among the members, the greater the knowledge sharing among the members and the confidence in their group [35]. This confidence drives the work efficacy to increase [36]. In other words, members of the organization continuously create new values by exchanging and combining knowledge to improve the performance of their group through shared recognition about their group [37]. Furthermore, as the members of the organization increasingly perceive that their group has sufficient competence and capabilities to achieve their goals, communication and cooperation among themselves will increase and their tasks would be successfully performed [36]. Therefore, this study established hypotheses as follows:

Hypothesis 6 Shared Vision will have a positive effect on Team-Efficacy.
Hypothesis 7 Trust will have a positive effect on Team-Efficacy.

2.2 *Knowledge Sharing*

Knowledge is a source of value creation for corporations and is recognized as an important business resource of companies that can overcome uncertain business environment and contribute to the company's competitive advantage and sustainable growth. In other words, knowledge is a combination of personal cognitive activities, experiences, and situations, and is useful information for problem solving and decision making [10, 11, 38]. Knowledge sharing is a process of fulfilling the so-called intellectual desire that is made by the interplay between knowledge sharing and actors who exchange knowledge. In other words, knowledge sharing is a process in which members exchange knowledge and create new knowledge through social networks [39]. Since successful knowledge sharing promotes shared intellectual capital, knowledge sharing is considered to be an important factor for corporate growth in the organization [11, 40].

Since knowledge expands when it is shared and its value increases too, achieving knowledge sharing depends on how effectively knowledge can be shared [41]. Knowledge sharing is an activity in which members of an organization share their knowledge to maximize the utilization of knowledge and thereby enhance the effectiveness of the organization [39]. Srivastava et al. [36] suggest that knowledge sharing and team-efficacy are highly related. As knowledge sharing increases, team-efficacy increases. Liao et al. [42] suggested that knowledge sharing is an activity involving methods for acquiring knowledge, and knowledge sharing as a unique characteristic, culture, or system of an organization affects organizational performance. Wang and Noe [43] suggested that knowledge sharing contributes to innovation and organizational competitiveness and the use of knowledge contributes to enhancement of organizational competitiveness. Therefore, knowledge sharing promotes work cooperation among the members of the organization, forming collective efficacy for the organization, and influencing the performance of the organization [36, 37, 43]. Therefore, this study established hypotheses as follows:

Hypothesis 8 Knowledge Sharing will have a positive effect on Team-Efficacy.
Hypothesis 9 Knowledge Sharing will have a positive effect on Team Performance.

2.3 *Team-Efficacy*

Efficacy refers to faith or beliefs that an individual's ability or ability to perform a particular task will lead to successful performance [33]. Team-efficacy is an extended sense of trust from individual to the team and refers to faith or beliefs the team members have toward skills and abilities that organize and perform needed for successful team performance [33]. In other words, team-efficacy is a shared perception as well as comprehensive assessment of the team's ability in performing a tasks by the team members themselves and viewed as an important asset that enhances team performance [20, 36, 44, 45]. High team-efficacy makes possible a more challenging and high-level goal setting, a deep involvement in team work, a high level of trust and solidarity, and a tendency to participate actively in the structuring and planning of tasks and in the adaptive process [46]. When team members perceive that their team has sufficient ability to accomplish the given tasks, they will develop a higher level of emotional attachment to the team.

Also, the team can use such sense of loyalty to make various efforts to enhance performance [19]. In addition, as the members of the team perceive that the team has sufficient competence and capability to achieve a given goal, the fair share of work is shared among the members to achieve the shared goals given to the team, and the level of communication and cooperation are increased [20]. In other words, team-efficacy plays a significant role in team members supporting and cooperating with each other and in forming positive and active interpersonal relationships, and helps improve team performance. Gully et al. [44] found a positive correlation between team-efficacy and team performance in a meta-analysis of team-efficacy. In addition, Stajkovic et al. [20] reported that organizational potentials positively affect group efficacy, and group efficacy has a positive effect on organizational performance. Liu et al. [47] suggested that the higher the collective efficacy of the new product development team, the more cooperative behaviors and joint decision-making are promoted, which has a positive effect on innovation performance. Therefore, this study established hypotheses as follows:

Hypothesis 10 Team-Efficacy will have a positive effect on Team Performance.

Based on the above hypotheses, study model in this study has been suggested as shown in Fig. 1.

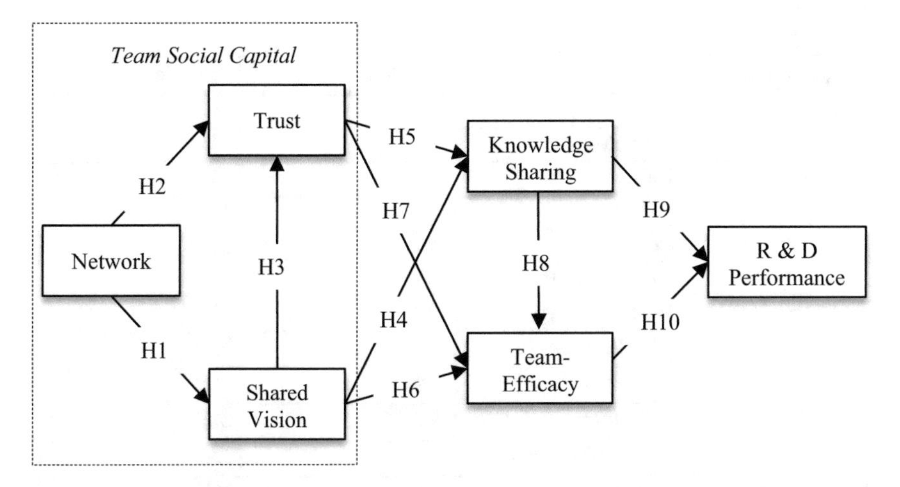

Fig. 1 Research model

3 Research Method

3.1 Samples and Data Collection

In order to verify the research model set up in this study, we collected data on companies that have been conducting or have experience in conducting R&D projects among domestic manufacturing companies. Based on the data provided by the SMBA, 2000 companies in the Korea, which will conduct business activities based on 2014, were selected as samples and surveyed using online e-mail for 4 months from October 2016 to February 2017 Respectively. A total of 230(11.5%) questionnaires were conducted to remove the questionnaires with missing or inadequate answers, and the final 219(11.0%) cases were selected as valid samples. Table 1 summarizes the sample of this study.

3.2 Measures

To ensure the content validity of the measurement tool, this study used the measurement items verified in the existing literature by revising and supplementing them according to the purpose of this study. First, the sub-dimensions of Social Capital of Team, which consist of Network, Shared Vision and Trust, were constructed into 4 items each in reference to the studies by Nahapiet and Ghoshal [1], Tsai and Ghoshal [22], Chen et al. [4] and Yu et al. [18], and were measured using 7-point Likert scale (Strongly disagree ~ Strongly agree). Also, Knowledge Sharing was constructed into 4 items in reference to the studies by Liao et al. [42], Wah et al. [48] and Tsai et al.

Table 1 Sample characteristics

Category and items		Sample size	Ratio (%)
Gender	Male	141	64.4
	Female	78	35.6
Age	20–29	29	13.2
	30–39	90	41.1
	40–49	67	30.6
	More than 50	33	15.1
Job career	Less than 5 yrs	34	15.5
	5–10 yrs	73	33.3
	10–15 yrs	59	26.9
	15–20 yrs	46	21.0
	More than 20 yrs	7	3.2
Project period	Less than 6 month	25	11.4
	6–12 month	82	37.4
	12–18 month	63	28.8
	18–24 month	25	11.4
	More than 24 month	24	11.0

[30] and was measured using a 7-point Likert scale. Team-Efficacy was constructed into 4 items in reference to the studies by Salanova et al. [49], Tasa et al. [37] and Emmerik et al. [50] and was measured using a 7-point Likert scale. Team Performance was constructed into 4 items in reference to the studies by Han and Hovav [16], Bardhan et al. [35] and Gu et al. [51] and was measured using a 7-point Likert scale. Table 2 summarizes the measured items and references of this study.

3.3 Analysis Method

For the analysis method and measurement tool of structural equation models, this study analyzed the results and verified the hypothesis using Amos 24.0. For the analysis of the structural equation model, the measurement model was estimated first, and then it was analyzed using the maximum likelihood that is widely used since the two-step approach that estimates the structural model, sample size and the normality assumption were found to be adequate.

Table 2 Confirmatory factor analysis base on reliability

Variable	Items of measurement	Factor L.D.	C.R.	Crb. Alpha
Network	Team members are intimate	.909	.894	.893
	Team members communicate effectively	.766		
	Team members share information	.741		
	Team members cooperate	.862		
Shared vision	Team members agreed with goals and vision	.792	.863	.793
	Team members agreed with work priority	.641		
	Share same hopes and vision	.639		
	Team members make efforts to achieve the team goal	.732		
Trust	Team members trust one another	.876	.888	.892
	Team members try to help one another	.843		
	Try to help other team members in difficulty	.755		
	Make efforts to achieve the team goal	.817		
Knowledge sharing	Share the knowledge about job	.883	.926	.921
	Share experience and knowledge about work	.852		
	Share knowledge about new work techniques	.874		
	Share new information	.844		
Team-efficacy	Confidence in performing work	.884	.939	.934
	Faith in achieving goals	.921		
	Find solution when a problem arises	.895		
	Faith that hardship can be overcome	.843		
R&D performance	Complete the work within schedule	.842	.934	.925
	Complete the work within budget	.886		
	Achieve the technical requirement	.864		
	Achieve the work requirement	.888		

4 Analysis and Results

4.1 Measurement Model

This study conducted confirmatory factor analysis to ensure the content validity of the measurement tool. For this, χ^2 = standard $\chi^2(\chi^2/\mathrm{df})$ RMSEA, TLI, CFI, and IFI were used to check goodness of fit. As a result, initial model did not exceed standard fitness threshold, so modified indices analysis were conducted, and measurement items that lowers unidimensionality were deleted (EP4). As a result of confirmatory factor analysis of modified measurement model, $\chi^2 = 550.223$ ($P =$ .000), $\chi^2(\chi^2/\mathrm{df}) = 2.337$, RMSEA = .078, TLI = .915, CFI = .927, IFI = .928, all indices suggested the measurement model used were fit. After verifying measurement model's fitness, reliability and validity were analyzed. For reliability, construct reliability should appear above .7, and average variance extract should be above 0.5. Additionally, for validity, two latent variables' AVE1 and AVE2 should bigger than squared value of its correlation. As a result of analysis, reliability and validity were verified and the detailed results are presented in Tables 2 and 3.

4.2 Structural Model

As measurement model's fitness, and reliability and validity of measurement items were verified, structural model analysis were conducted. As a result of structural model's fitness test, $\chi^2 = 63.051$ ($p = .000$), $\chi^2(\chi^2/\mathrm{df}) = 2.327$ was above threshold 3, and RMSEA = .078 was below standard of 0.08. Moreover, RMSEA = .078, TLI = .916, CFI = .926, IFI = .927 all of indices appeared above recommended value of 0.9 and therefore, the structural model' goodness of fit of the research model was verified.

Table 3 Discriminant validity

Variable	1	2	3	4	5	6
1. Network	**.681**[a]					
2. Shared vision	.389	**.614**[a]				
3. Trust	.498	.518	**.664**[a]			
4. Knowledge sharing	.448	.594	.582	**.759**[a]		
5. Team efficacy	.384	.415	.462	.534	**.795**[a]	
6. R&D performance	.275	.364	.361	.469	.476	**.781**[a]

[a]*AVE* Average Variance Extract

4.3 Hypotheses Test

After structural model's fitness was confirmed, research hypotheses were tested. As a result, first, for Team Social Capital's structural relationship Network appeared to have an effect on Shared Vision, $\beta = .440$ (C.R. = 8.663, $p = .000$), thus, supporting H1. Also, Network appeared to have an effect on Trust, $\beta = .473$ (C.R. = 5.543, $p = .000$), thus, supporting H2. Shared Vision appeared to have an effect on Trust, $\beta = .694$ (C.R. = 5.117, $p = .000$), thus, supporting H3. Second, for relationship between Team Social Capital and Knowledge Sharing, both Shared Vision, $\beta = .701$ (C.R. = 5.180, $p = .000$), and Trust, $\beta = .409$ (C.R. = 5.166, $p = .000$) had positive effect on Knowledge Sharing, therefore, H4 and H5 were supported. Third, for relationship between Team Social Capital and Team-Efficacy, Shared Vision had a significant effect on Team-Efficacy, $\beta = .200$ (C.R. = 1.257, $p = .209$), while Trust had an effect on Team-Efficacy, $\beta = .255$ (C.R. = 2.883, $p = .004$), thus H7 was supported while H6 was not supported. Fourth, for relationship between Knowledge Sharing and Team-Efficacy, Knowledge Sharing appeared to have positive effect on Team-Efficacy, $\beta = .403$ (C.R. = 3.708, $p = .000$), thus, supporting H8. Fifth, for relationship between Knowledge Sharing and R&D Performance, Knowledge Sharing appeared to have positive effect on Team Performance, $\beta = .354$ (C.R. = 4.659, $p = .000$), thus, supporting H9. Lastly, for relationship between Team-Efficacy and R&D Performance, Team-Efficacy had a positive effect on Team Performance, $\beta = .367$ (C.R. = 4.755, $p = .000$), thus H10 was supported. The results of hypotheses test are summarized in Table 4 and Fig. 2.

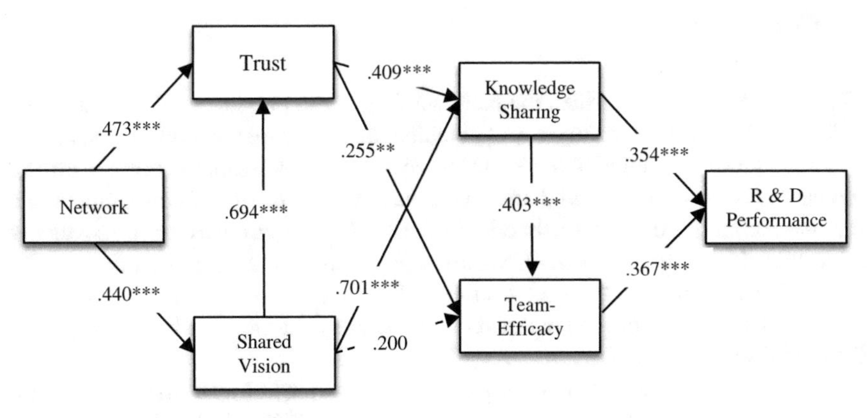

Fig. 2 Research model and path efficiency

Table 4 Discriminant validity

Hypotheses	Path	Estimate (β)	Std. Error	C.R. (t)	Supported/not supported
H1	Network → Shared vision	.440	.051	8.663***	Supported
H2	Network → Trust	.473	.085	5.543***	Supported
H3	Shared vision → Trust	.694	.136	5.117***	Supported
H4	Shared vision → Knowledge sharing	.701	.135	5.180***	Supported
H5	Trust → Knowledge sharing	.409	.079	5.166***	Supported
H6	Shared vision → Team-efficacy	.200	.159	1.257	Not Supported
H7	Trust → Team-efficacy	.255	.089	2.883**	Supported
H8	Knowledge sharing → Team-efficacy	.403	.109	3.708***	Supported
H9	Knowledge sharing → R&D performance	.354	.076	4.659***	Supported
H10	Team-efficacy → R&D performance	.367	.077	4.755***	Supported

*: $p < .05$, **: $p < .01$, ***: $p < .001$

5 Conclusions

This study empirically verified the factors affecting the project team's performance in the R&D project environment and obtained the following meaningful results.

First, as a social capital of the R&D project team, the network has positive effects on shared vision and trust, and shared vision has an effect on trust. These results show that the formation of social networks in the R&D project environment positively affects the formation of common values, goals, and trust among the members. In other words, it can be understood that the R&D project team maintains various types of networks, thereby forming the trust relationship by sharing the vision and goal of the organization [27, 28].

Second, the shared vision and trust among R&D project team members have a positive effect on knowledge sharing. This result shows that in R&D project environment, team members share the vision and values of organization and individual. It can be said that it induces communication and promotes the sharing of knowledge among members by recognizing the value of exchange and combination of knowledge. In other words, it can be understood that the formation of high trust among team members enables members to participate actively in the social exchange process and exchange and utilize knowledge [18, 29, 30].

Third, trust between R&D project team members affects positively team-efficacy, and trust formation between team members positively affects organizational development by promoting joint efforts and cooperative behaviors. In other words, in the IT project environment, the members form a consensus among them through social capital and establish a high trust relationship, so that the members share fair work load, and the communication and cooperation increase [20, 32]. On the other hand, shared vision has a significant effect on team-efficacy. It can be stated that, to promote innovation in R&D project environment, trust between team members should be formed and mutual understanding and cooperation activities should be preceded [50].

Fourth, knowledge sharing among R&D project team members positively affects team-efficacy and team performance. These results show that knowledge sharing among team members in the R&D project environment is an important factor for enhancing the performance by effectively promoting team competitiveness and confidence that they can successfully perform tasks by promoting cooperation among members [36, 37, 43].

Finally, team-efficacy in the R&D project has a positive effect on team performance. As the team members perceive that their team has sufficient competence and capability to achieve their goals, it promotes cooperative behavior and joint decision making. Also, it has positive effect on performance [20].

The purpose of this study is to identify the success factors of the R&D project from the viewpoint of the social capital which is emerging as the core resource of enterprise management today. To this end, social capital, knowledge sharing, and team-efficacy were identified as key drivers for project team performance in the R&D project environment, and empirical verification of the relationship between variables was conducted. It is meaningful to look at the structure and role of social capital, which has been unexplained in the achievement of the project performance by the project executing organization in the R&D project environment where the importance is increasing recently. It is also meaningful to see that the social capital formed between the members of the project performing organization strengthens organizational capacity and eventually leads to performance.

This study confirms the importance of building team capital and strengthening team competence in the R&D project environment and suggests the strategic direction for the success of R&D projects and the enhancement of team competitiveness.

References

1. Nahapiet, J., Ghoshal, S.: Social capital, intellectual capital, and the organizational advantage. Acad. Manag. Rev. **23**(2), 242–266 (1998)
2. Adler, P.S., Kwon, S.W.: Social capital: prospects for a new concept. Acad. Manag. Rev. **27**(1), 17–40 (2002)
3. Kor, Y.Y.: Direct and interaction effects of top management team and board compositions on R&D investment strategy. Strateg. Manag. J. **27**(11), 1081–1099 (2006)

4. Chen, M.H., Chang, Y.C., Hung, S.C.: Social capital and creativity in R&D project teams. R&d Manage. **38**(1), 21–34 (2008)
5. Thomas, V.J., Sharma, S., Jain, S.K.: Using patents and publications to assess R&D efficiency in the states of the USA. World Pat. Inf. **33**(1), 4–10 (2011)
6. Brettel, M., Mauer, R., Engelen, A., Küpper, D.: Corporate effectuation: entrepreneurial action and its impact on R&D project performance. J. Bus. Ventur. **27**(2), 167–184 (2012)
7. Ebrahim, N.A.: Virtual R&D teams: a new model for product development. Int. J. Innov. **3**(2), 1–27 (2015)
8. Côté, S., Healy, T.: The well-being of nations: the role of human and social capital. Organisation for Economic Co-operation and Development, Paris (2001)
9. Hoegl, M., Weinkauf, K., Gemuenden, H.G.: Interteam coordination, project commitment, and teamwork in multiteam R&D projects: a longitudinal study. Organ. Sci. **15**(1), 38–55 (2004)
10. Huang, C.C.: Knowledge sharing and group cohesiveness on performance: an empirical study of technology R&D teams in Taiwan. Technovation **29**(11), 786–797 (2009)
11. Tang, C., Ye, L.: Diversified knowledge, R&D team centrality and radical creativity. Creativity and Innov. Manage. **24**(1), 123–135 (2015)
12. Chen, Y., Vanhaverbeke, W., Du, J.: The interaction between internal R&D and different types of external knowledge sourcing: an empirical study of chinese innovative firms. R&D Manage. **46**, 1006–1023 (2016)
13. Khedhaouria, A., Montani, F., Thurik, R.: Time pressure and team member creativity within R&D projects: the role of learning orientation and knowledge sourcing. Int. J. Project Manage. **35**(6), 942–954 (2017)
14. Liu, Y., Keller, R.T., Shih, H.A.: The impact of team-member exchange, differentiation, team commitment, and knowledge sharing on R&D project team performance. R&D Manage. **41**(3), 274–287 (2011)
15. Gu, Q., Wang, G.G., Wang, L.: Social capital and innovation in R&D teams: the mediating roles of psychological safety and learning from mistakes. R&D Manage. **43**(2), 89–102 (2013)
16. Han, J., Hovav, A.: To bridge or to bond? Diverse social connections in an IS project team. Int. J. Project Manage. **31**(3), 378–390 (2013)
17. Vincenzo, F., Mascia, D.: Social capital in project-based organizations: its role, structure, and impact on project performance. Int. J. Project Manage. **30**(1), 5–14 (2012)
18. Yu, Y., Hao, J.X., Dong, X.Y., Khalifa, M.: A multilevel model for effects of social capital and knowledge sharing in knowledge-intensive work teams. Int. J. Inf. Manage. **33**(5), 780–790 (2013)
19. DeRue, D.S., Ashford, S.J.: Who will lead and who will follow? A social process of leadership identity construction in organizations. Acad. Manag. Rev. **35**(4), 627–647 (2010)
20. Stajkovic, A.D., Lee, D., Nyberg, A.J.: Collective efficacy, group potency, and group performance: meta-analyses of their relationships, and test of a mediation model. J. Appl. Psychol. **94**(3), 814–828 (2009)
21. Bartsch, V., Ebers, M., Maurer, I.: Learning in project-based organizations: the role of project teams' social capital for overcoming barriers to learning. Int. J. Project Manage. **31**(2), 239–251 (2013)
22. Tsai, W., Ghoshal, S.: Social capital and value creation: the role of intrafirm networks. Acad. Manag. J. **41**(4), 464–376 (1998)
23. Chow, W.S., Chan, L.S.: Social network, social trust and shared goals in organizational knowledge sharing. Inf. Manag. **45**(7), 458–465 (2008)
24. Castro, I., Roldán, J.L.: A mediation model between dimensions of social capital. Int. Bus. Rev. **22**(6), 1034–1050 (2013)
25. Coleman, J.S.: Social capital in the creation of human capital. Am. J. Sociol. **94**, 95–120 (1988)
26. McAllister, D.J.: Affect-and cognition-based trust as foundations for interpersonal cooperation in organizations. Acad. Manag. J. **38**(1), 24–59 (1995)

27. Lee, J., Park, J.G., Lee, S.: Raising team social capital with knowledge and communication in information systems development projects. Int. J. Project Manage. **33**(4), 797–807 (2015)
28. Gulati, R., Sytch, M.: Does familiarity breed trust? Revisiting the antecedents of trust. Manag. Decis. Econ. **29**(2–3), 165–190 (2008)
29. Chang, H.H., Chuang, S.S.: Social capital and individual motivations on knowledge sharing: Participant involvement as a moderator. Inf. Manag. **48**(1), 9–18 (2011)
30. Tsai, Y.H., Ma, H.C., Lin, C.P., Chiu, C.K., Chen, S.C.: Group social capital in virtual teaming contexts: a moderating role of positive affective tone in knowledge sharing. Technol. Forecast. Soc. Chang. **86**, 13–20 (2014)
31. Pinheiro, M.L., Serôdio, P., Pinho, J.C., Lucas, C.: The role of social capital towards resource sharing in collaborative R&D projects: evidences from the 7th framework programme. Int. J. Project Manage. **34**(8), 1519–1536 (2016)
32. Molina-Morales, F.X., Martínez-Fernández, M.T.: Social networks: effects of social capital on firm innovation. J. Small Bus. Manage. **48**(2), 258–279 (2010)
33. Bandura, A.: Editorial. Am. J. Health Promotion **12**(1), 8–10 (1997)
34. Meng, X.: The effect of relationship management on project performance in construction. Int. J. Project Manage. **30**(2), 188–198 (2012)
35. Bardhan, I., Krishnan, V.V., Lin, S.: Team dispersion, information technology, and project performance. Prod. Oper. Manage. **22**(6), 1478–1493 (2013)
36. Srivastava, A., Bartol, K.M., Locke, E.A.: Empowering leadership in management teams: effects on knowledge sharing, efficacy, and performance. Acad. Manag. J. **49**(6), 1239–1251 (2006)
37. Tasa, K., Taggar, S., Seijts, G.H.: The development of collective efficacy in teams: a multilevel and longitudinal perspective. J. Appl. Psychol. **92**(1), 17–27 (2007)
38. Makhija, M.: Comparing the resource-based and market-based views of the firm: empirical evidence from Czech privatization. Strateg. Manag. J. **24**(5), 433–451 (2003)
39. Hackbarth, G.: The impact of organizational memory on IT systems. AMCIS 1998 Proc. 197 (1998)
40. Hu, L., Randel, A.E.: Knowledge sharing in teams: social capital, extrinsic incentives, and team innovation. Group Org. Manage. **39**(2), 213–243 (2014)
41. Liebowitz, J.: Knowledge management and its link to artificial intelligence. Expert Syst. Appl. **20**(1), 1–6 (2001)
42. Liao, S.H., Fei, W.C., Chen, C.C.: Knowledge sharing, absorptive capacity, and innovation capability: an empirical study of Taiwan's knowledge-intensive industries. J. Inform. Sci. **33**(3), 340–359 (2007)
43. Wang, S., Noe, R.A.: Knowledge sharing: a review and directions for future research. Hum. Resour. Manage. Rev. **20**(2), 115–131 (2010)
44. Gully, S.M., Incalcaterra, K.A., Joshi, A., Beaubien, J.M.: A meta-analysis of team-efficacy, potency, and performance: interdependence and level of analysis as moderators of observed relationships. J. Appl. Psychol. **87**(5), 819 (2002)
45. Mathieu, J., Maynard, M.T., Rapp, T., Gilson, L.: Team effectiveness 1997–2007: a review of recent advancements and a glimpse into the future. J. Manag. **34**(3), 410–476 (2008)
46. DeRue, D.S., Hollenbeck, J., Ilgen, D., Feltz, D.: Efficacy dispersion in teams: moving beyond agreement and aggregation. Pers. Psychol. **63**(1), 1–40 (2010)
47. Liu, J., Chen, J., Tao, Y.: Innovation performance in new product development teams in China's technology ventures: the role of behavioral integration dimensions and collective efficacy. J. Prod. Innov. Manag. **32**(1), 29–44 (2015)
48. Wah, C.Y., Menkhoff, T., Loh, B., Evers, H.D.: Social capital and knowledge sharing in knowledge-based organizations: an empirical study. Int. J. Knowl. Manage. **3**(1), 29–48 (2009)
49. Salanova, M., Llorens, S., Cifre, E., Martínez, I.M., Schaufeli, W.B.: Perceived collective efficacy, subjective well-being and task performance among electronic work groups an experimental study. Small Group Res. **34**(1), 43–73 (2003)

50. Emmerik, H., Jawahar, I.M., Schreurs, B., De Cuyper, N.: Social capital, team efficacy and team potency: the mediating role of team learning behaviors. Career Dev. Int. **16**(1), 82–99 (2011)
51. Gu, V.C., Hoffman, J.J., Cao, Q., Schniederjans, M.J.: The effects of organizational culture and environmental pressures on IT project performance: a moderation perspective. Int. J. Project Manage. **32**(7), 1170–1181 (2014)

Study on Detection Algorithm of Live Animal in Self-bag-Drop Kiosk in Airport Using UWB Radar

Kiwon Jung, Younghwan Bang and Sun-Myung Hwang

Abstract The study is conducted detection by UWB (Ultra-Wide Band) to prevent against safety accidents which could be occurred in Self Bag Drop installed and unmanned operated in airport by the unexpected intrusions such as alive animals, humans especially. To conduct the study, we applied algorithm with the detected signal of an object movement and respiratory cycle during standstill by filtering/signal processing S/W. For the proper study, we installed and operated UWB radar in the mock-up frame of Self Bag Drop. We study the operation of the system by the detected alive object's status in the Self Bag Drop and analyse its doppler pulse.

Keywords UWB · Self bag drop · Airport

1 Introduction

In compliance with increasing the number of airport passengers globally, the most of airports intend to present Self Check in or Self Bag Drop. These kinds of self service units are extremely beneficial to reduce passenger's operation time, cost etc. Recently, Incheon International airport terminal 1 and 2 introduced Anti-Intrusion Self Bag Drop which is more secured and safe and its type of units are being introduced in other airports as well (Fig. 1).

The feature of anti-intrusion Self Bag Drop is closeable type composed with safety door, weighing & conveyor belts to deliver the baggage to the plane eventually.

K. Jung (✉)
SCom CNS Inc, Daejeon, Republic of Korea
e-mail: kjung@scomcns.net

Y. Bang
Korea Institute of Industrial Technology, Chungcheongnam-do, Republic of Korea
e-mail: bangyh@kitech.re.kr

S.-M. Hwang
Daejeon University, Daejeon, Republic of Korea
e-mail: sunhwang@dju.kr

R. Lee (ed.), *Software Engineering Research, Management and Applications*, Studies in Computational Intelligence 789, https://doi.org/10.1007/978-3-319-98881-8_7

Fig. 1 Self bag drop at Incheon Airport (closed type)

However, the Self Bag Drop has difficulty to detect exact object identification such as a child or animal when baggage weighed under unmanned operation. It obviously makes the safety accident due to fail of the detection in the Self Bag Drop. There are various detection system such as video or thermal method against those targets mentioned above but they are not fully and exactly operated detecting the targets. In the case of thermal imaging analysis, the each unit's imaging pattern has less reliability due to surrounded clothes or other obstacles on the objects and obviously expensive of the detector. Therefore, competitive and reliable detecting method with high performance is necessary to prevent any safety accident by the alive object intrusions.

UWB (Ultra-Wide Band) is the Bandwidth radar technology with high spatial resolution, low power technology and reasonable cost, which is specially divided in 3–10 GHz. The permissible frequency range in each country is different but 3–7 GHz or 7–10 GHz are particularly permitted for UWB radar. Compared with current Bio-signal detection technology, UWB has high frequency range for its signal and high resolution so that UWB is more effective to detect target's fine movements. For its advantage, UWB is studied in the various field of industries like military, medical, especially for bio-signal and non-contact detection.

The biggest benefit of UWB technology application is detectable in the condition of non-invasive, non-contact, non-intrusive with low power consuming. It also available to detect the alive object regardless of its matter of conscious. The vibrant and active research in the field of medical is Polysomnography by applying UWB. But the general PSG monitors respiratory cycle during sleep by the contacted sensors on the body (Fig. 2).

There are various devices applied for Polysomnography (PSG) measurement [1]. However, they all performed the function with contacting sensors on the body, which are inconvenience for the test. To overcome those all complicated process, UWB is

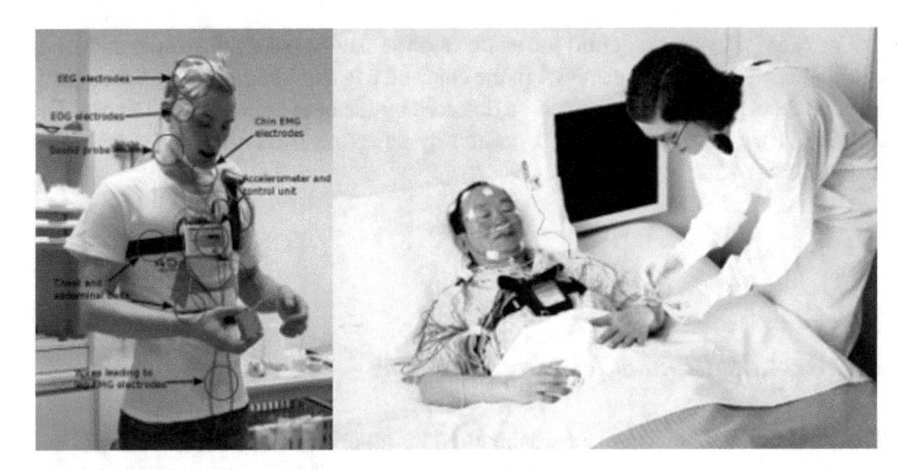

Fig. 2 PSG measurement device

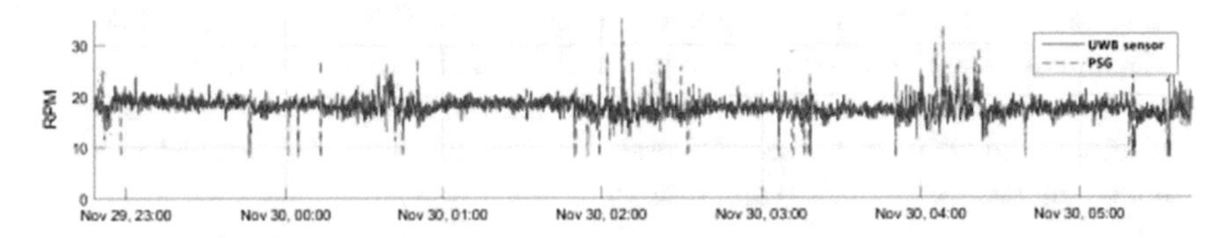

Fig. 3 Comparison of respiratory cycle monitoring result between PSG and UWB [1]

being studied. Not available to recover whole devices but it's alternative for both 2 belts contacting to chest and abdominal.

This study was carried out by the Ministry of Science and Technology's support for research funds provided by the SW compute industry source technology

As per the Fig. 3, UWB's respiratory cycle measurement result shows almost same as PSGs. According to data from the manufacture of UWB (Novelda) [2], it shows error rate between minimum 1% to maximum 3%. Compared with current PSG's complicated measurement, the interest in new measurement is rising along with UWB.

As well as bio-signal detection research such as blood pressure, cardiac impulse, has been actively increased for the specific medical application of usage with UWB's enhanced technology. Nevertheless, UWB mostly is applied for users' desired purpose of usage with their own development but signal processing commonly performs by applied Novelda's SoC. In this study, we suggest the research result, analysis data and algorithm from the detection of the target's (human) movement and its respiratory cycle during standstill in the inner space of Self Bag Drop through Novelda UWB radar and signal processing S/W.

In this study, the kinetic factors that can induce the Doppler effect that can distinguish live animals can be classified with movement and respiration.

For example, cases of child intrusion into the device, user falls down and loses consciousness, putting live animal in the checked bag and internal components damage by wheeled bag's moving can be detected by the above-mentioned factors. The primary target is for accidents prevention by detecting live animal movement and respiration.

2 Detection Experiments

2.1 Test Environmental Configurations

The experimental environment was applied by mock-up frame same with self bag drop device installed at Incheon International Airport. The equipment used is shown in Fig. 4.

The size of the section requiring internal detection is 580 mm in width, 1000 mm in height and 1200 mm in depth. The required detectable-distance should be longer than 1000 mm and have a radiation width covering width of 580 mm. So the antenna used in this study has a horizontal/vertical radiation angle of about 65° and has radiation characteristics as shown in Fig. 5.

In this test, radar is installed at a distance of 30–60 cm from the target in mock-up frame similar to of belt space inside self bag drop device. Human respiration detecting measurement were operated in other opened space separated from self bag drop device. For verification about measurement distance and accuracy, the respiration period tests were also operated.

Fig. 4 Self bag drop device angle

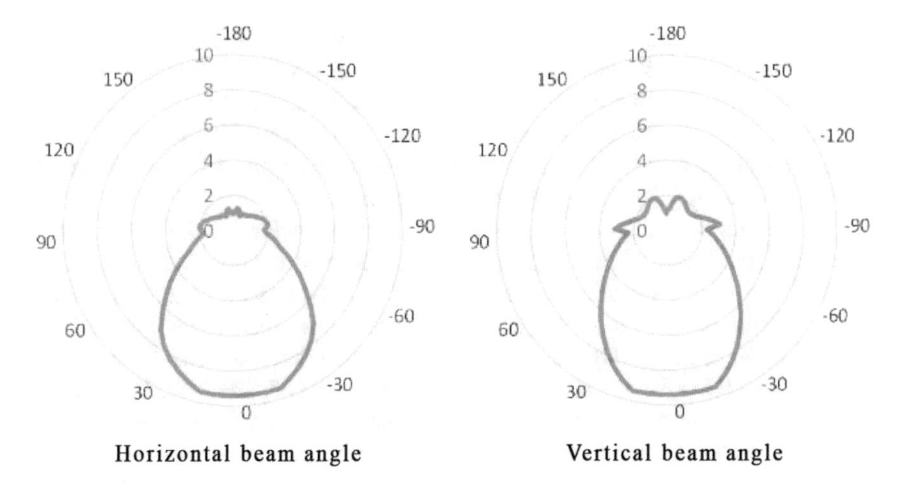

Fig. 5 UWB beam angle

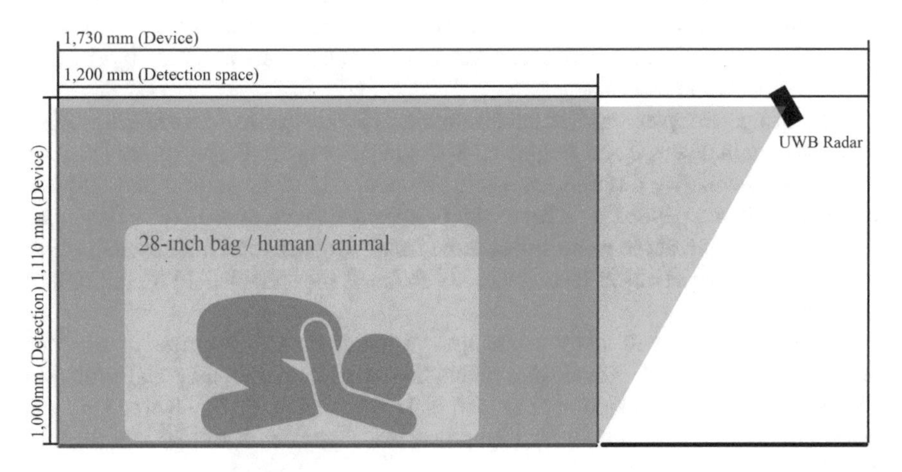

Fig. 6 Experimental configurations

The experiment was conducted after specifying the installation location to reflect the product performance. For live animal detection experiments, test subject generate movement in the mock-up frame, and a 28 in. luggage was used for non-alive object's movement detection (Fig. 6).

2.2 Test Environment Configurations

In the above-mentioned kinematic factor, the variation between the transmitting and receiving ultra wide band radio for movement and respiration can be deduced by the doppler equation. Variation by breathing periodic chest volume changes by inhalation and exhalation which has smaller than body movements. In contrast to breathing, body movements are not periodic. But it has larger variation than respiration. Therefore, when the variation is small and periodicity is observed, it can be judged that the breathing is detected in the state without movement. Or the variation is large and has not periodicity, it can be judged jus movement.

The steps for detecting live animal in a checked baggage using the doppler radar signal are; Transmitting UWB signal; Receiving the reflected doppler; Determining whether there is the variation of reflected doppler; Generating emergency alarm when variation is equal to or larger than a predetermined set value; Determining whether a change in the value has regular period when the amount of change is less than a defined set value; Comparing the detected amount of variation with a defined reparation pattern when the variation has regular period; Generating emergency alarm when the detected amount of variation is determined to respiration pattern.

The signal processing process according to the above conditions is defined as an algorithm. 17 frames per second are set, and two doppler region matrices classified in slow/fast are executed in parallel. The slow movements are analyzed by the data of the last 20 s of the radar frame and the fast movements are analyzed by the data of the last 6 s of the radar frame.

Both Range-Doppler matrices have individual Noise Maps to determine if a reflection at a certain distance and frequency is above a threshold. Creating and enabling a Noise Map will give different threshold values at different distances and frequencies and is recommended to achieve the best performance. If the Noise Maps are disabled a fixed threshold value will be used for all distances and frequencies, which in most cases will result in less sensitivity to small movements and higher risk of false detections.

The Noise Maps will adapt to changes in the environment unless Noise Map Adaptation is disabled. Noise Map adaptation works continuously and will over time remove presence detection of reflectors that are stationary. The Noise Map will not adapt if a still person with breathing frequency between 8 and 30 Respirations Per Minute (RPM) is detected.

The Fast Range-Doppler matrix with its Fast Movement Detector will detect presence quickly, typically when a person enters the Detection Zone. The Fast Movement Detector has two states, Movement or No Movement. The Fast M/N Combiner uses these states to determine the LocalState Fast. An M/N Combiner determines that M out of N detections need to be a certain value for the output to change [3] (Fig. 7).

Detection is performed by separating fast motion (LocalState Fast) and slow motion (LocalState Slow) based on the reflected doppler. The respiration detecting is included in the slow motion.

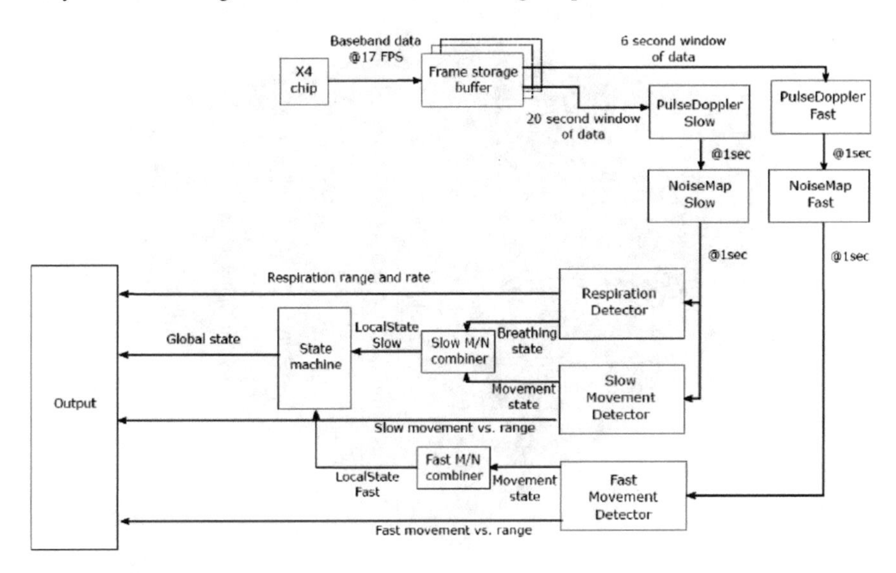

Fig. 7 Respiration profile signal processing diagram [3]

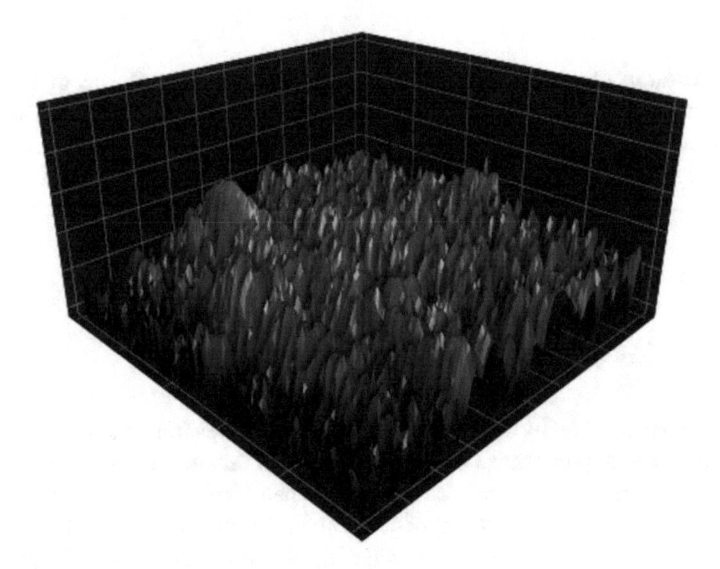

Fig. 8 Doppler pulse of non-detection

2.3 Movement Object Detection Test

An reflected doppler was detected by internal instruments even though no motion was detected, but doppler echo waves were receiving a low variation of less than −10 dB (Fig. 8).

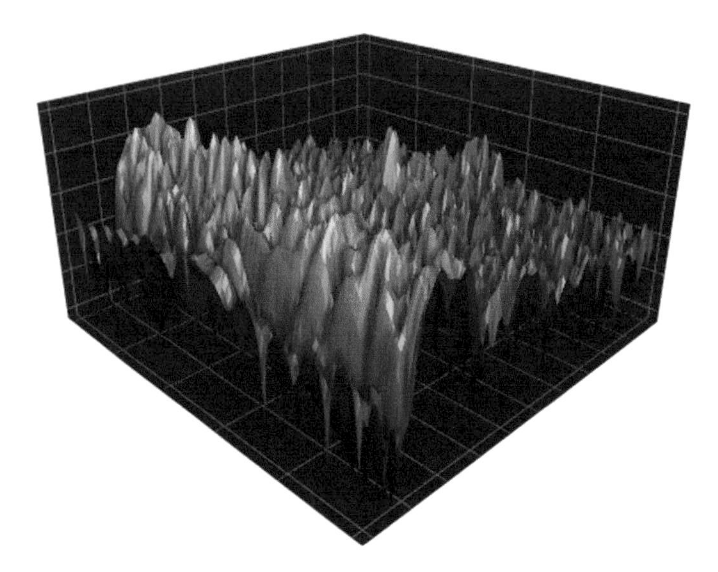

Fig. 9 Doppler pulse of movement detection

As a result of the reflected doppler detection test, moving objects of a size less than 10 cm can also be detected. The detectable movement displacement was able to detect even the smallest movement less than 1 cm. Even with a small displacement in the variation, the reflected doppler differs by more than 5 dB, so the data can be clearly distinguished between those that detected the fuselage and those that did not (Fig. 9).

2.4 *Bio-signal (Respiration) Detection Test*

Respiration is detection by displacement changes in the resting person's chest movement. When the doppler pulse due to the large and fast motion becomes stable, it will begin to detect live animal's respiration. When some movement was detected, it was shown as a different amount depending on the reflection characteristic of the subjects, and the signal intensity could be detected according to the doppler pulse detection characteristic. And it also detects multiple movements along the distance (Fig. 10).

As a result of the reflected doppler detection test, moving objects of a size less than 10 cm can also be detected. The detectable movement displacement was able to detect even the smallest movement less than 1 cm. Even with a small displacement in the variation, the reflected doppler differs by more than 5 dB, so the data can be clearly distinguished between those that detected the fuselage and those that did not (Fig. 11).

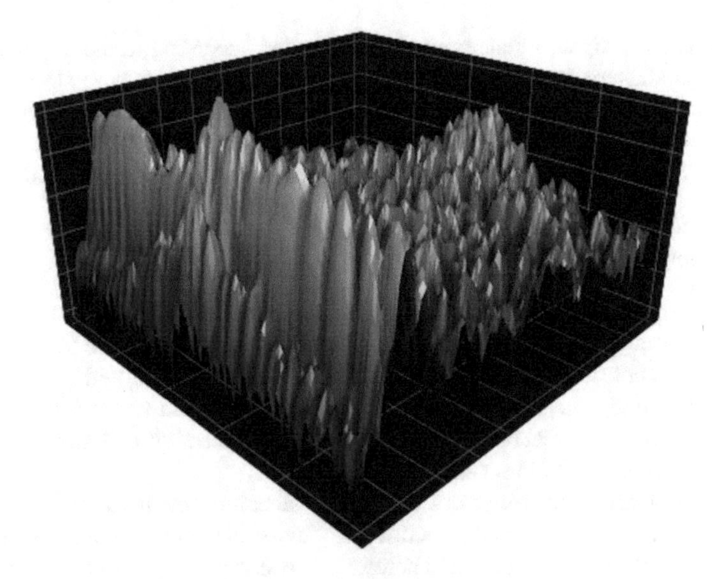

Fig. 10 Doppler pulse of general movement detection

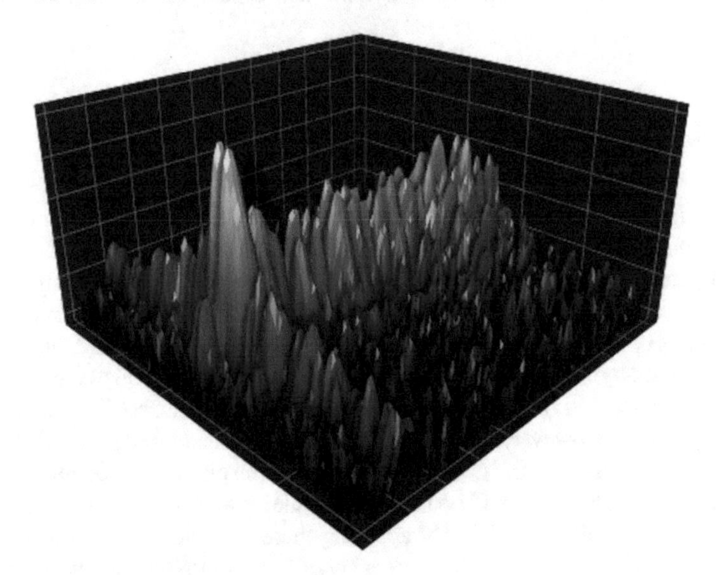

Fig. 11 Doppler pulse of respiration detection

Respiration is detection by displacement changes in the resting person's chest movement. When the doppler pulse due to the large and fast motion becomes stable, it will begin to detect live animal's respiration. When some movement was detected, it was shown as a different amount depending on the reflection characteristic of the subjects, and the signal intensity could be detected according to the doppler pulse detection characteristic. And it also detects multiple movements along the distance.

To analyze the characteristics of doppler during the detection of the movement and the respiration and check accuracies, the respiration period at the time of breath detection was detected and the doppler characteristics were analyzed (Table 1).

- 10 subjects
- Measuring distance 1, 3, 5 m (sitting, stopped)

To analyze the characteristics of doppler during the detection of the movement and the respiration and check accuracies, the respiration period at the time of breath detection was detected and the doppler characteristics were analyzed.

When breathing was detected, the amount of displacement was less than 0.4 Hz, and the distance data received from the radar communication unit was almost the same as the distance of the set subject.

Respiration detection is interrupted when any other movement occurs. However, in the case of unconscious human detection, only bio-signals (breathing and heartbeat) are generated without movement. Therefore, it is expected that accurate breathing detection will not affect the operating scenario with parameters that can detect living things.

3 Conclusion

The study has found that small movements of objects within the measurement range can be sensed and that the characteristics of Doppler pulses varied according to the characteristics of the movements and the displacement of the movements. The respiration detection have detected the respiration by detecting a difference in the amount of very small changes of below ±0.4 Hz of changes in the Doppler pulse which is caused by the displacement difference of the sternum due to the inspiration and the exhalation and in case that no major movement occurs, continuous detection has been possible. However, due to the fact that the high resolving power that detects movements less than 1 cm. it may cause loss of respiration detection when the movement occurs, but there will be no major problems with the scenario of operation of the baggage handling device. By applying these detection properties, we could devise a method for detection and processing of live animals as shown in Fig. 12.

By analyzing the Doppler reverberation, Scenarios aim at immediately inducing a response at the site by raising an alert with unconscious live animal detection when detecting the movement and respiration according to the characteristics of the movement.

In addition, it is considered that the additional biometrics detection technologies may be necessary by investigating the possible exceptional situation. If, for instance, it is possible to detect small movements such as heartbeats or blood flow, then it can be used as additional indicators that can be cross-validate.

Table 1 Respiration test result

Subject		1	2	3	4	5	6	7	8	9	10
Detecting (m)		O	O	O	O	O	O	O	O	O	O
RPM	1	17.2	18.2	16.4	16.6	16.9	17.3	17.6	16.3	17.7	19.9
	3	17.4	17.5	16.7	16.8	16.5	16.8	22.5	17.9	16.8	20.3
	5	17.1	17.9	19.8	16.8	17.8	21.6	18.0	17.9	18.4	20.2

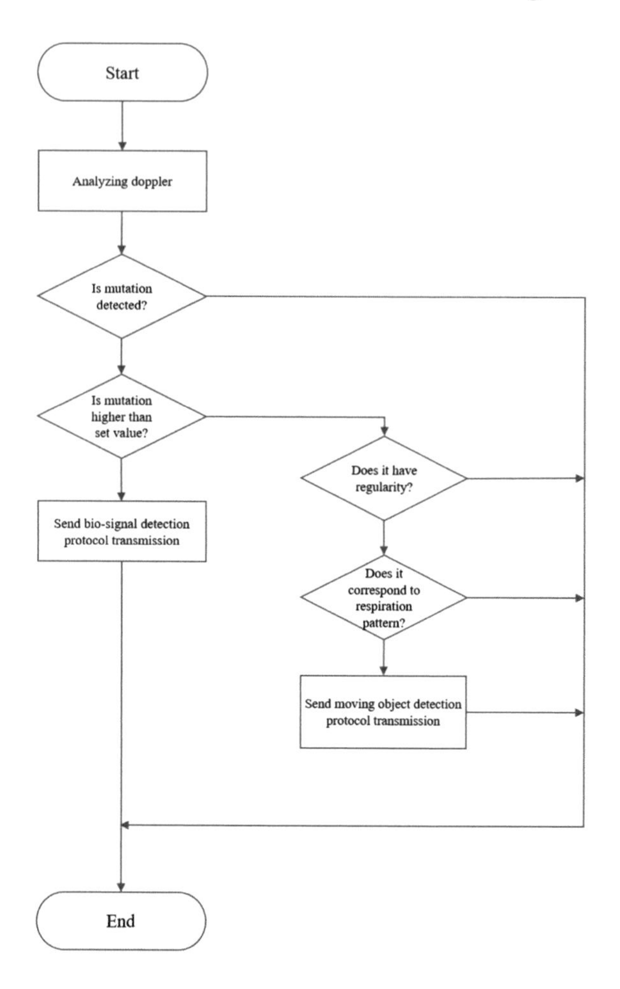

Fig. 12 Bio-signal detection processing flow chart

Moreover, the trend of unmanned operation through automation of facilities utilized in a variety of areas, as well as self-back-drops at airports, which were mainly targeted, began for a long time. As it is still in its early stages, these changes have been carried out with the most priority of the convenience and cost reduction rather than safety. Technology has been developed focusing on operation and automation in unmanned environments, but it is necessary to prepare a preventive measure concerning accident that may occur to smoothly operate equipment and secure the user safety in unmanned environments. The complex live animal detection system that utilizes UWB radar is capable of distributing various types of unmanned equipment such as self-backed drop in airports, as well as BHS (Baggage Handling System), cargo automation processing system etc. Moreover, it is expected to be utilized as a preparation technique against various types of safety accidents in various fields.

Acknowledgements This study was carried out by the Ministry of Science and Technology (ICT)'s support for research funds provided by the SW compute industry source technology. [Standard interface-based commercial service platforms and auto handling with multiple features (Issuance of boarding pass and consignment of baggage) for building advanced smart airport terminals. Development by KIOSK].

References

1. Novelda, A.S.: XeThru Sleep Monitoring—A Polysomnography (PSG) Comparative Study (2018, February)
2. Novelda, A.S.: High-end and Non-contact Sensor Technology for Respiration (2014)
3. Novelda, A.S.: X4M200 Datasheet Respiration Sensor (2017, September)

A Study on Success Factors for Business Model Innovation in the 4th Industrial Revolution

Sung-Hwan Yoon, Nguyen Si Thin, Vo Thi Thanh Thao, Eun-Tak Im and Gwang-Yong Gim

Abstract World businesses are being governed by unicorn enterprises based on creative business models. An analysis will be made to see what factors are important for unicorn enterprises. While the existing studies emphasized regulations and entrepreneurship aspects, the unicorn enterprises that are currently governing the world have been realized by having innovative business models as the key competence with entrepreneurship and regulations added. Consequently, it will be studied by using ANP technique what an important factor the understanding on innovation of business models is.

Keywords Unicorn · Business model · Business model innovation
Hyper connectivity · Creative innovation

1 Introduction

Unicorns as startups exceeding a trillion in value are emerging in more than one per quarter since 2011, in more than one per month since 2014, and in more than 70 ea. in one year of 2016 [1]. Collision between two worlds of reality and virtuality is giving birth to global gigantic unicorns such as Uber and Airbnb. This is just a

S.-H. Yoon · N. S. Thin · V. T. T. Thao · E.-T. Im · G.-Y. Gim (✉)
Soongsil University, Seoul, Korea
e-mail: gygim@ssu.ac.kr

S.-H. Yoon
e-mail: ryan4ari@gmail.com

N. S. Thin
e-mail: thinns168@gmail.com

V. T. T. Thao
e-mail: thaovo90dn@gmail.com

E.-T. Im
e-mail: iet030507@gmail.com

R. Lee (ed.), *Software Engineering Research, Management and Applications*, Studies in Computational Intelligence 789, https://doi.org/10.1007/978-3-319-98881-8_8

representative revolutionary phenomenon of the 4th industrial revolution. Online to offline (O2O) convergence between virtuality and reality is being displayed as the unicorn, which is the key icon of the 4th industrial revolution [2]. In addition to such enterprises as Airbnb and Uber, O2O optimization that provides offline with optimized service after online analysis of optimization using artificial intelligence is diffused into all areas of human life such as hospital, plant, travel, etc. causing a boom of unicorns. They converge and circulate offline reality and online virtuality to create optimized values of prediction and customization [3].

Considering the list of unicorn enterprises published by CB Insights in March 2017, Korean businesses such as Coupang (25th), Yellowmobile (31st), and CJ Games (69th) have been enlisted. However, in the case of Coupang and Yellow mobile, they continue to experience difficulties. Indeed, the number of unicorns has risen exponentially in recent years, but a lot of them (Square, Dropbox, Evernote…) also was winding down and facing with stiff competition (Apple, Microsoft, Google, Amazon…) [4]. In addition, the 4th industrial revolution is disrupting almost every industry in every country. The successful firm needs a success business model or business model innovation that gained an increasing amount of attention in management research and among practitioners. The emerging the success of business model literature addresses an important phenomenon but lacks theoretical underpinning, and empirical inquiry is not cumulative. Business model is assumed as the structure, content and governance of transactions inside the firm as well as between the firm and its external partners in supporting the firm's creation, delivery as well as capture of value [5–7]. The first, business models reflect the strategic choices of the company [6, 8], the choice of creative innovation requires that the company defines those ways to create, deliver and capture value in conjunction with external partners that are consistent with open innovation [9, 10]. Empirical evidence strongly proposes that organizational design, practices and capabilities need to be aligned with creative innovations, so as to positively affect the sourcing of knowledge from external parties and its subsequent exploitation for innovation [11–14].

The second, the 4th industrial revolution facilitates the surging of collective intelligence, collaborative governance, and the resolution of complicated, global problems. We will change to an era of unprecedented wealth, prosperity, and opportunity for all. Either way, opportunities will emerge for awareness and tenacious entrepreneurs to create business models that integrate the natural creative abilities of humans with the operative efficiencies of machines, resulting in new ways of living, working and exploring the world around us. It is the hyper connectivity connecting to social networks and other sources of information. Furthermore, within hyperconnected contexts, firms do not only cooperate and collaborate but also attempt to achieve advantage [15, 16].

As aforementioned, unicorns are emerging and very successful in U.S.A and China, but it is the humble in Korea. Further, there is little research that clearly joins business models to open creative innovation and hyper connectivity. Little is known about how firms need to design their business models to match different among them. So, this paper concern what factors should be considered important to give birth to unicorn enterprises, it will be studied herein accordingly on what an important factor the understanding on business model innovation in the 4th industrial revolution by using ANP technique.

2 Literature Review

2.1 Business Model

The business model represents "the content, structure, and governance of transactions designed so as to create value through the exploitation of business opportunities" [8, 9]. Based on the fact that activities are connected by transactions, the definition of company's business model is conceptualized as "a system of interdependent activities that transcends the focal firm and spans its boundaries" [17].

However, recent advances of the Industry 4.0 in communication and information technologies, such as the emergence as well as rapid expansion of Internet and the speedy drop in communication and computing costs, have allowed the advancement of new methods for creating and delivering value, which have offered expansion for creating unconventional exchange mechanisms, transaction architectures [18] as well as highlighting the capacities for the design of new boundary-spanning organizational forms [19, 20].

Indeed, such developments have opened new horizons for business models' design by facilitating companies to change the way they organize and participate in economic exchanges fundamentally, both within as well as across company and industry boundaries [21]. This includes the ways in which firms interact with suppliers as well as with customers [22].

Technological innovation is acclaimed in most advanced societies; that is an attractive and natural reflection of the values of a technologically developing society. However, the creation of new organizational forms, organizational methods (like the moving assembly line). Whilst such innovation may seem less heroic to many citizens even many scientists and engineers without it technological innovation may be bereft of reward for pioneering individuals, as well as for pioneering enterprises and nations. The capacity of a company either nation to catch value will be compromised deeply unless the capacity exists to design new business models. The development business model innovation from business model is necessary in the new era. So what the business model innovation is, we will explain in the next section.

2.2 Business Model Innovation

Business models as well as business model innovation have received significant attention not only in literature but also industry and business model innovation is increasingly suggested as a key element to the success of business [23, 24]. Business model innovation occurs when companies improve their existing business models or introduce the new ones. Business model innovation is a concept based on the principle that companies innovate by leveraging their internal resources and capabilities [7]. Various viewpoints on business model were presented in the literature. According to Teece, business model articulates how the company will convert capabilities and resources into economic value [25]. Business model is defined as a completed description on how a company does business [7]. Based on a widespread range of literature, Richardson proposed a combined view of the components of a business models included value proposition, value creation, delivery system, and value capture system [26]. More particularly, Osterwalder and his partner described a business model as a series of essential factors: the value proposition, activities, resources, partners, distribution channels, cost structure and revenue model [27]. There are differences among academics about what establishes a business model, there is a comprehensive consensus around four critical constituents [28]: value proposition [22, 25, 29], value creation [22, 30], value delivery and value capture [27, 31]. A fifth constituent, value communication, is also considered as an important feature of a business model [32]. These five elements of a business model permit researchers to define as well as specify a business in how it creates and appropriates value. A business is only created when a company matches it product/service to a set of customers [33]. Five constituents and their respective sub-constituents are synthesized in the Fig. 1 below [28].

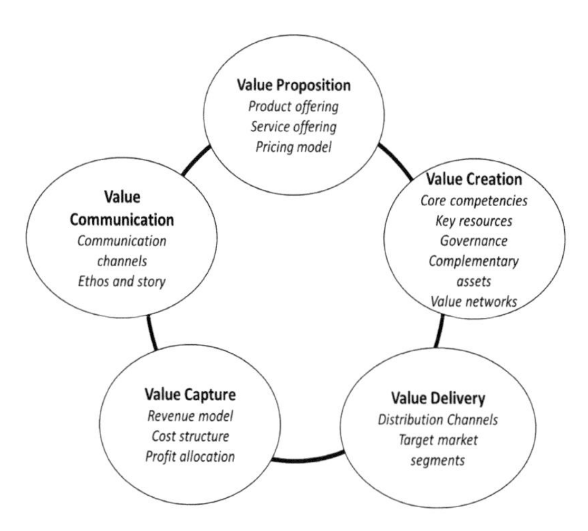

Fig. 1 Key components of business model innovation [28]

2.3 Unicorns and Successful Business Models in the 4th Industry Revolution

The economic impact of the 4th industrial revolution is expected to result in heightened economic inequality. It is the era of digital economy, decentralized everything, and meta business model. The world is about to go through the greatest transformation in its history [34]. We can see that new business models have emerged in the 4th industrial revolution that distinguish themselves from "the way business was conducted" in the 3rd industrial revolution. These new business models are rooted at the network of the absolute changes in technology (connectivity, computing power, and automation) and a generational or societal change.

Unicorn companies such as Uber, Airbnb, Alibaba… are the great examples of the many companies that are currently changing the business world as one knows it. These new platform business models enable companies to do business differently and are a source of value creation in the digital economy. It is true that Uber has brought a revolution, not just as a taxi company, but also as a business model where businesses reach out to serve customers at their location. A lot of startups have already made their app like Uber and many others have made small iterations to launch startups in different industry verticals. Alibaba's business model is complex when it comes to the products that it has all its platforms combined. Thus only commissions on sales would not work well for the revenue generation of the company and that is where the entire ecosystem works together to cater all types of customers, Alibaba, TaoBao, Aliexpress, TMall… are all working together to service to large and small businesses and individual customers.

Eventually, we can assume that a common theme that has been applied to these new business models is hyper connectivity and creative innovation, of which more details will be given.

2.4 The Success Factors for Business Model Innovation

Business model innovation is considered as the key element to maintain the competitive advantage of firms. It would be difficult to find any startups in general as well as unicorn companies in particular that do not make a revolutionary business model. Examining which factors is important for innovating business model in 4th industrial revolution is necessary for not only academics but also practitioners. Our study aimed to provide two clusters of factors that affect to the successful innovation of business model include: Hyper connectivity and Creative Innovation.

2.4.1 Hyper Connectivity (HC)

A new world of connectivity is emerging. People and the things around us are increasingly connected to networks and sharing information, collaborating more across borders and gaining incredibly rich insights from big data. Advances in the 4th industrial revolution's digital technologies have increased the capacities for out sourcing business activities to crowds of independent contributors. It's a one-side performance of hyper connectivity referring to harnessing the power of a huge amount of people to solve a hard problem as a group [35].

Using the collective intelligence of a crowd extends a new scope of business opportunities. For instance, Amazon's Mechanical Turk (http://www.mturk.com/) is a crowdsourcing market place run by Amazon.com. In this market place, businesses can demand human intelligence tasks (tasks which computers have difficulties to do or cannot execute, such as selecting a pizza store amongst different photos, identifying music in CDs, etc.) and the Mechanical Turk brings a workforce that can finish HITs for a monetary fee, usually very low [36].

There is a burgeoning potential for the integration of collective intelligence processes which broaden beyond the companies' traditional boundaries in the way they adapt to deliver value that can be translated in new business models. Indeed, the company's new business models made feasible by hyperconnected processes will determine and influence the way if the organization will succeed, and therefore, "an unexceptional innovation with a superior business model may prove more profitable than a superior innovation with an unexceptional business model" [37].

Hyperconnected world is creating the potential to enhance the value chain of corporation, since it serves the same goal as outsourcing, but reduces the financial risks, rises flexibility, allows extensive access to talent as well as ideas, lowers costs, rises capabilities and cuts time down to market. Hence, there are many examples of organizations using crowdsourcing effectively to either create or derive value from markets by using innovative products and services based on collective input.

In summary, crowdsourcing and open innovation can make use of the same business model, in which they both apply the hyper connectivity system for innovation and problem solving. In the present paper, we consider two more factors network effect and platform in this model.

Crowdsourcing (CR)

The term was coined by Jeff Howe with the aim to "describe a new Web-based business model that harnesses the creative solutions of a distributed network of individuals through what amounts to an open call for proposals" [38]. Crowdsourcing is a combination between Crowd and Outsourcing which alluding to the out-sourcing of collective activities to crowd an independent mass of people [39].

One of the more frightful advantages of crowdsourcing is a job would be ended up doing better by customers then the paid professionals would do in the first place. Crowdsourcing is taken into a new dimension with the support of Internet, which gives much free and better access to the crowd. According to Neto and partners, crowdsourcing has become a well-known methodology for bringing ideas from outer sources together, collecting as well as filtering input through web-based tools [40]. Industry 4.0 is about connectivity that includes the integration with an ecosystem of a customer such as customers and suppliers. The integration layer for driving joint innovation may be provided by Internet of things and relevant platforms. In addition, consumers as well as customers might be integrated via crowdsourcing as well. Because data and information are critical for Industry 4.0, crowdsourcing will introduce new advantages such as a new way to collect data collaboratively and then it enables which will make industrial processes easier [41].

Open Innovation (OP)

The complicated environment forces firms to move from classical approaches of creating value (closed innovation strategy) and capturing value (closed business model) to a new open approach (open business model and open innovation strategy) [42]. Open innovation is related to the systematical encouragement and exploration of a wide range of inner as well as outer sources for innovation opportunities, which consciously with companies' resources as well as capabilities, and broaden exploitation of those opportunities via multiple channels [43]. Open innovation supposes inner ideas can also be brought out the market via outer channels outside the firm's current business to generate additional value [44]. There are many studies identifying advantages of open innovation such as leveraging outer knowledge inputs to advance inner innovations and broaden the markets for outer use of innovation. New business models can be identified by an adaptive business model in an open innovation platform [45]. In industry 4.0, major trends are shorter service and product lifecycles as well as the need to accelerate time to market. Therefore, it will lead to the need for innovation and introduce both complication and cost. However, it also brings firms to their limit of innovation capacity as well as internal capabilities. Therefore, to master the innovation game as well as to stay competitive in rapid-changing market, open innovation is the best strategy [46]. We live in hyper-connected world where platforms and open innovation play as key elements for identifying the challenges of companies more easily. That is the reason why open innovation was chosen as important factor in hyper connectivity cluster for successful business model in the industry 4.0.

Network Effects (NET)

Network effect is a phenomenon in which current customers of any product or service can benefit in some way whenever such product or service is used and adopted by additional customers. This effect is generated by many customers and value is added to their product's usage. The speed of network effect then rises with co-creation exponentially where the customers subscribe automatically for the additional value that they are getting. In the booming era of Internet, network effects play an important role in the success or failure of marketplace platform business. By rising customer base, market share as well as the overall value proposition of product then creating increased profits, network effects help scale business. Nevertheless, in order to harness the network effect's power and scale business, designing an effective business model consistent with a way its platform creates differentiated and unique value for each set of customers is very important. The more subscribers the platform gains, the more the network effect comes into action. In the hyper connected world, a well-functioning network effect enhances the collective intelligence of crowds, being why network effect was chosen as an important factor for successful business model in industry 4.0 era.

Platform (PL)

The term "platform" is considered as one of the most misunderstood terms that we use nowadays. A platform represents architecture—a design for services, products and infrastructure expediting the interactions among network users—plus a set of rules such as the rights, protocols and pricing terms that take control transactions. A platform business model grows by the absolute presence of its catalytic force—the customer, does not by its own virtue. Platforms, which provide, attract the building blocks and means that bring all of these together. In the 4th industrial revolution era, it generated a shift in the way the world traditionally worked, a new shape to business models was created—the platform business model. The last years have seen the new web platforms' development which providing the essential infrastructure for collective creation. Well-known examples of such platforms are Kluster, Cambrian House, or CrowdSpirit and this phenomenon is designated by the concept of "crowdsourcing" [47]. With the development of network and digitalization in 4.0 industrial revolution, the economy becomes the place where number of connections in development, manufacture and sales, both nationally as well as globally is increasing. Platforms associate all the relevant stakeholders as well as enable them to communicate. Besides, the combination between crowdsourcing and platform is considered as necessary strategy to collect ideas for innovation.

2.4.2 Creative Innovation (CI)

With unicorn company, the ideal summed up in the phrase "Creative innovation" have become a powerful part of business thinking. Innovations are actually original prerequisite of competitiveness [48]. The 4th industry revolution forced most of businesses to have to savings in all business fields. Otherwise, it should be noted that the 4th industry revolution for will gone and come again revival the economy and re-distribution of markets. Successful companies will be the ones that will have applied a creative innovation, will invest in research, development and innovate. Innovation is concern as a key business process; it means that companies are trying through them to gain a competitive opportuneness. The elementary precondition for the creation and adoption of innovation in the company is well-formulated and implemented creative innovation.

The variety of reaching for creating innovation conduct the fact that creating innovation as a system of work with innovation in the enterprise is evolving. It is possible to form a relatively universal model that will provide successful implementation and follow-up realization of creative innovations in the company. When using a model of creative innovative must be suitably chosen structural elements of the model to show to the importance of innovation in the enterprise [49].

Research on creative innovation suggests that companies benefit differentially from adopting open innovation strategies; however, it is inexplicit why this is so. One possible explanation is that companies' business models are not adapted to open strategies. There are not many researches that links business models to creative innovation explicitly. That is, while such innovations distinct significantly with regard to, for instance the number and classes of actors involved [50, 51] and the innovation process's phases that are kept open with regard to interact with outside knowledge sources [48, 52]. How firms need to design their business models to match different creative innovation is little to known about. To answer this question, we review relevant literatures on creative innovation and business models and subsequently propose four factors of creative innovation affecting the success business. They are: less is more, long tail, personalization, disruptive technology.

Long Tail (LO)

The characteristics of this approach are offering low volume products in large range. Firms active in many niche markets are suitable to adopt this approach. The long tail company is concentrated in attracting consumers buying niche products [35]. According to Kroik, the overall profitability of long tail companies may be better than companies following the separation concept [53]. It is easy to understand because niches markets offer companies the chances to obtain greater profit margins, which make large total profits when companies manufacture a large assortment of products. Such business models are found to be well adapted to niche clusters of clients as well as maintaining good relations with them [53]. As a result, Internet plays an important role in long tail firms. Existing companies have focused on products/services that attract consumers buying the most to make money. However, in the 4th industrial revolution, based on Big Data, information exchange has become easier, and Big Data can create invisible demand. This allows a variety of items to be handled by a variety of minorities. Therefore, long tail is an important factor in creative innovation.

Less-Is-More Innovation (LE)

Less-is-more is a kind of offering clients less of what many of them expect, it is a simpler or stripped-down product/service. Less of what some clients perceive as a good thing may really be good for some clients, and very good for the bottom line of producer. These companies stripped off some attributes that clients expect, simplifying the product in the process. In several cases, firms combined existing features in innovative ways that make the new product simplified. In other words, they also added new features that may not have been expected but making sure the final product was cheaper and simplified than older ones in the market [50]. In the world of excess everything, it is realized that less is the best, anything confusing, wasteful or excessive should be removed intelligently or never added in the first place. The innovators know that providing a meaningful and memorable experience depends on user engagement through a subtractive approach is best-achieved [54].

Disruptive Technology (DI)

Disruptive technology is defined as either a new technology or new combination of existing technologies whose application to new commercialization challenges or problem areas can create entirely new major technology product paradigm or cause major shifts ones [55]. Through the introduction of products/services, which are effectively cheaper, better and more convenient, disruptive technology create growth in the industries in which they penetrate or even create new industries. Disruptive technologies present a revolutionary change in the conduct of operations or processes [56]. It is suggested that disruptive technologies might enter as well as expand emerging market niches, improve time by time and eventually attack existing products in their traditional markets. Therefore, disruptive technology brings huge challenges to incumbents because the changes it brings to the profit models as well as existing value networks of companies. In the era of Industry 4.0, computers and automation come together entirely in a new way along with robotics connected to computer systems remotely equipped with machine learning algorithms. Therefore, disruptive technology plays an important role for making the competitive advantage of firms, being why disruptive technology is listed as one of the factors affect to the successful business model. The convergence between industries, the convergence between disciplines provides a new business perspective.

Personalization (PE)

For any industries, understanding customers as well as their demands is considered as the main challenge. In the 4th industrial revolution era, we are leaving the traditional business model of offering standard product/service selections behind and entering a new business model in which the purchase process is ruled by personalization. Personalized product/service offerings help firms attract new customers via network effects, platforms. Through personalization, firms also give existing customers a reason to make purchases repeatedly by improving the products' value perception.

Moreover, personalization allows firms to collect customers' information that can be used to establish, launch new products in the future, this is possibly the biggest advantage. In personalization, customers are empowered and gone further than lip service as well as the noise around crowdsourcing. Generally, it is about listening to customers as well as providing tailored customer service. Companies, along with the support of technology offer product/service, which is personalized to customers' individual and immediate needs at competitive prices, and personalization is considered as the key factor for transformative business model [49]. In the Industry 4.0 era, products need to be designed and produced in a smart way to meet requirements of mass customization, personalization and flexible smart manufacturing, being the reason personalization should be considered as an important factor for the success business model in Industry 4.0.

2.5 AHP (Analytic Hierarchy Process)

AHP is a decision-making technique which is developed by Thomas L. Saaty in 1980. AHP helps selecting optimal alternatives under various qualitative and quantitative criteria. It first hierarchizes a given decision-making problem. A pairwise comparison matrix $a(n \times n)$ is created with the relative weight or significance of each factor/criterion in the lower stratum which is right below the upper level. Weight is calculated with eigenvectors [57].

$$A\omega = \lambda_{max}\omega \tag{1}$$

In which:

A—Pair wise comparison matrix
ω—Eigenvector
λ_{max}—Maximum eigenvector

In AHP perspective, the intuition, enriched experience, and such of a decision maker are important. Therefore, it has the merit of handling qualitative information easily which is hard to measure but needed to be considered for decision-making, and quantitative information [57].

In AHP, there are two important considerations in drawing an expert group's knowledge: consistency ratio (CR) and integration of expert knowledge.

CR is the tool to check the reliability of expert knowledge. According to Thomas L. Saaty, λmax-n is an effective tool to measure consistency since it always has an equal value to or bigger value than n a positive transposed matrix and only when matrix A is consistent, the value can be n [58]. Using this principle, we can get consistency index (CI).

$$CI = \mu = \lambda_{max} - n/n - 1 \tag{2}$$

In which:

λ_{max}—Maximum eigenvector
n—The number of factors

When CI is divided by random index (RI), it yields consistency ratio (CR). Thomas L. Saaty maintained that when consistency is perfect, CR is equal to 0. On the contrary, when consistency is not perfect or poor, CR is bigger than 0. Therefore, when $CR > 0.1$, it is necessary to make a decision again or modify it [51]. However, some social science fields accept it up to 0.2 (20%) for allowance because it is not easy to establish independence between the upper and lower criteria, considering the characteristics of questionnaire items [59].

There are two ways to aggregate the experts' opinions: Arithmetic mean and geometric mean. Woo Chun-sik and his partners stated in their comparative study of bankruptcy prediction model that geometric mean should be used for qualitative information to aggregate expert opinions and it has a higher prediction rate of bankruptcy than when arithmetic mean is used. Therefore, geometric mean can aggregate expert opinions more firmly [57].

To design a model to apply to the given decision-making problem, using AHP, 4 steps should be followed as seen in Fig. 2 [60].

Step 1: Make a decision hierarchy with strata (or levels) (environmental scenarios, criteria, and alternatives to implement) for a decision-making problem set on a top objective. That is, it is a job to hierarchize the criteria for evaluation and the factors in a lower stratum should be described more specifically and in detail. However, too many attributes in one level produce too many for relative comparison. Therefore, it is desirable that the number of criteria in one level should be no more than 9 [61].
Step 2: Collect the data to compare pairwise with each criterion (factor) for the environmental scenario and the alternatives for the criterion. Make AHP matrix with the pairwise comparison data by level (stratum). That is, once a decision hierarchy is formed, make a pairwise comparison matrix to compare the factors with each other in each level. In general, the scale used for pairwise comparison is from 1 to 9.
Step 3: To evaluate the relative weights of decision-making factors, solve the eigenvalues of the matrix constructed in Step 2. Estimate the relative weights of decision-making factors and verify CR (Consistency Ratio) to measure the reliability of the respondent' professionality. In general, a relative weight is obtained by calculating importance using Saaty (1980)'s eigenvector. Only when CR is less than 0.1, we consider that reliability is secured [60].
Step 4: Aggregate the relative weights of the decision-making factors to prioritize the factors in each level. Then the priority of the items to evaluate is decided. It provides the basic information for resource allocation and alternative choice.

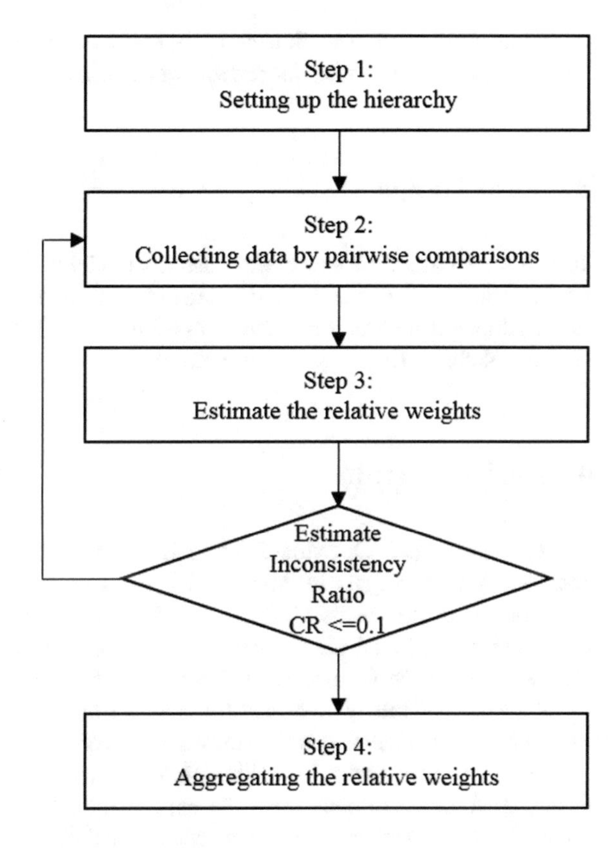

Fig. 2 The design procedure of one AHP model [60]

2.6 ANP (Analytic Network Process)

AHP has the demerit that it cannot reflect the inter-dependency of various factors in a real-life decision making. To compensate it, Thomas L. Saaty developed ANP technique from improving AHP. ANP is a decision-making methodology that considers the internal/external dependency among a target, criteria, and alternatives or the inter-dependency of the factors derived from the process of analysis in gathering experts' opinions and analyzes the correlation between those factors, on which a network structure is formed and their relative importance (priority) is drawn through pairwise comparison [62].

Using Super-matrix that expresses the weights of the nodes (factors) and the clusters (the groups of nodes) into a matrix, ANP comes by relative importance. It multiplies Cluster-matrix—calculated through the comparison of the clusters—by Unweighted Super-matrix indefinitely, which is made with the pairwise comparison date of the factors (nodes) in a cluster. Then, it results in Limiting Super-Matrix on which the relative weights of considerations are finally judged.

Unlike AHP, ANP uses Super-matrix to draw weight. Therefore, it is relatively more complicated and takes more time for calculation, which is the demerit of ANP.

3 Study Design

To find the factors for a successful business model in the 4th Industrial Revolution, an interrelation (dependency and independence) matrix was built with the success factors drawn from literature review, as seen in Fig. 3 and established ANP model. And Fig. 4 shows the designed research model.

4 Analysis Result

In this paper, Super Decision 3.0 and Excel were used to calculate and analyze consistency and weights in ANP model. Because this research is an exploratory study on the determinants of data transaction, it was judged that it was necessary to be lenient on CR: Data consistency was approved if CR is less than 0.2. The questionnaire where CR exceeds 0.2 was excluded from weight analysis.

Of 12 respondents on a successful business model, 6 showed CR below 0.2, so importance was calculated with their responses on pairwise comparison.

Thomas L. Saaty said that arithmetic mean and geometric mean could be used to aggregate the importance that the experts thought of [63]. When it was determined that the experts involving in a decision-making process possess a high level of expertise, weighted arithmetic mean could be used after considering their weights. Geometric means are calculated by geometrically averaging out the values of the equal components (entries) of a pairwise comparison matrix. With them, a new gen-

			Hyper Connectivity				Creative Innovation			
		Code	a-1	a-2	a-3	a-4	b-1	b-2	b-3	b-4
Hyper Connectivity	Crowdsourcing	a-1		1	1	1		1	1	1
	Open innovation	a-2	1							1
	Network effect	a-3	1			1		1		
	Platform	a-4	1	1	1		1		1	
Creative Innovation	Less-is-more	b-1				1				
	Long tail	b-2	1		1	1			1	1
	Personalization	b-3		1		1		1		1
	Disruptive technology	b-4	1	1	1		1	1	1	

Fig. 3 Factor interrelation matrix

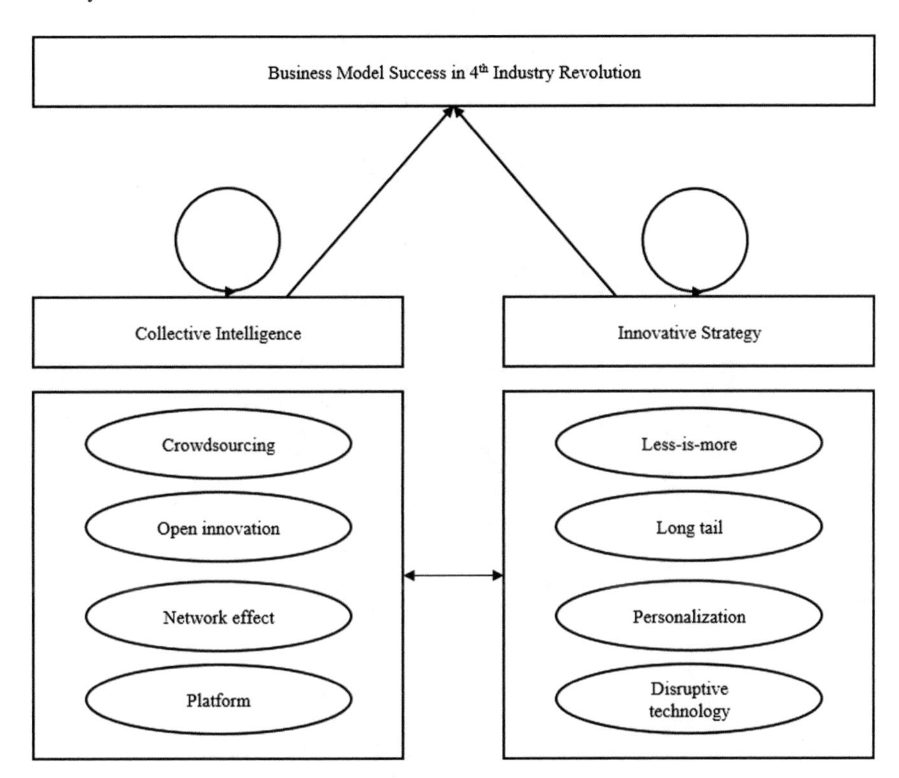

Fig. 4 Research model

Table 1 Cluster matrix

	HC	CI
HC	0.197967	0.455913
CI	0.802033	0.544087

eral matrix can be constructed and weight can be obtained from this matrix. It has been proved that geometric mean is better in terms of the accuracy and consistency of a model in case the opinions of various experts are reflected at once [62] (Table 1).

Unweighted Super-matrix was formed with the pairwise comparison data of the factors (nodes) in a cluster (Table 2).

The final relative importance of the considerations was decided on Weighted Super-matrix (resulting from the multiplication of Super-matrix by a reference cluster weight) and Limiting Super-matrix (resulting from the indefinite multiplication of Weighted Super-matrix) (Tables 3 and 4).

In the cluster of Innovative Strategy, Disruptive Technology (0.228558) turned out to be the most important factor, followed by Long Tail (0.207354) and Personalization (0.194628) in the order (Table 5).

Table 2 Unweighted matrix

		CI				IS			
		CR	OP	NET	PL	LE	LO	PE	DI
HC	CR	0	0.3289	0.1880	0.2451	0	0.3239	0.3437	0.3503
	OP	0.2841	0	0	0	0	0	0	0.6496
	NET	0.2339	0.6710	0	0.7548	0	0.6760	0	0
	PL	0.4818	0	0.8119	0	1	0	0.6562	0
CI	LE	0	0	0	0.1011	0	0	0	0
	LO	0.40197	0	0.45527	0.2974	0	0	0.5933	0.4409
	PE	0	0.3714	0	0.6013	0	0.5432	0	0.5590
	DI	0.59802	0.6284	0.54472	0	1	0.4567	0.4066	0

Table 3 Weighted matrix

		HC				CI			
		CR	OP	NET	PL	LE	LO	PE	DI
HC	CR	0	0.0651	0.0372	0.04851	0	0.1476	0.1567	0.1597
	OP	0.0562	0	0	0	0	0	0	0.2962
	NET	0.0463	0.1328	0	0.1494	0	0.3082	0	0
	PL	0.0953	0	0.1607	0	0.4559	0	0.2991	0
CI	LE	0	0	0	0.0811	0	0	0	0
	LO	0.3223	0	0.3651	0.2385	0	0	0.3228	0.2399
	PE	0	0.2979	0	0.4823	0	0.2955	0	0.3041
	DI	0.4796	0.5041	0.4368	0	0.5440	0.2485	0.2212	0

Table 4 Limit matrix

		HC				CI			
		CR	OP	NET	PL	LE	LO	PE	DI
HC	CR	0.1100	0.1100	0.1100	0.1100	0.1100	0.1100	0.1100	0.1100
	OP	0.0738	0.0738	0.0738	0.0738	0.0738	0.0738	0.0738	0.0738
	NET	0.0917	0.0917	0.0917	0.0917	0.0917	0.0917	0.0917	0.0917
	PL	0.0866	0.0866	0.0866	0.0866	0.0866	0.0866	0.0866	0.0866
CI	LE	0.0070	0.0070	0.0070	0.0070	0.0070	0.0070	0.0070	0.0070
	LO	0.2073	0.2073	0.2073	0.2073	0.2073	0.2073	0.2073	0.2073
	PE	0.1946	0.1946	0.1946	0.1946	0.1946	0.1946	0.1946	0.1946
	DI	0.2285	0.2285	0.2285	0.2285	0.2285	0.2285	0.2285	0.2285

Table 5 Research result

Cluster	Name	Weight	Rank
HC	CR	0.110064	4
	OP	0.073892	7
	NET	0.09178	5
	PL	0.08669	6
CI	LE	0.007034	8
	LO	0.207354	2
	PE	0.194628	3
	DI	0.228558	1

5 Conclusion

Judging from the general life cycle of startups, most of them disappear from the ecology in 5 years, not overcoming 'the valley of death'. Only a few break through a breakeven point using an original and creative business model and settle down on a stable growth. This situation is not going to be much different in the era of the 4th Industrial Revolution. Only the factors of a business model will be re-shaped and rearranged in a direction suitable for the survival in the waves of the era. The successful business model will follow the new trends of personalized production and service rather than for all and knowledge platform will shift from close to open frame. Furthermore, products will be simplified and minimalized because consumers feel tired of a complicate social stricture.

In this study, a literature review and a survey with an expert group was carried out to explore the important factors that compose an innovative business model for the 4th Industrial Revolution. Based on the results, variables were grouped into variables (Hyper Connectivity and Creative Innovation) and the sub-factors for each variable were designed; Hyper Connectivity has under it (1) crowdsourcing, (2) open innovation, (3) network effect, and (4) platform while Creative Innovation has (1) less is more, (2) long tail, (3) personalization, and (4) disruptive technology. I obtained a total of 8 factors to which ANP analysis was applied to get their relative importance (priority). Here is the importance from most to least: (1st) disruptive technology > (2nd) long tail > (3rd) personalization > (4th) crowdsourcing > (5th) network effect > (6th) platform > (7th) open innovation > and (8th) less is more.

The most important factor for an innovative business model is 'Disruptive Technology'. As the 4th Industrial Revolution fuses existing technologies with new ones to create a paradigm shift and eventually produce new products and services, it is very natural that disruptive technology is emphasized. In other words, disruptive technology means 'innovation' itself. Therefore, it is very inclusive (of all other factors) and thus the expert group considers it as the most important.

Followed in the order were 'Long Tail' and 'Personalization'. As it turned over to the 21st century, production paradigm changed from mass production and sale to small production and sale, focusing on personalized products and services. Long tail and personalization are the concepts developed at that time, but still valid up to date and will be more important in the future. Technical/technological development (e.g., A.I. and 3D printing) enables companies to offer customer-unique products and services. Customer's behavior patterns and preferences are analyzed for the recommendation of the best suitable and differentiated products from others. If we say the market has been divided into segments for management, it will be more than 10,000 segments from now on, which are micro-segment markets, and customers are offered new experiences. Therefore, it must be true that long tail and personalization are more important concepts for those markets to come.

Next important factors are crowdsourcing, network effect, platform, and open innovation. They all represent collective intelligence. When groups gather in a common platform, they tend to create the value that cannot be created when they disperse in different platforms or space. Like other factors mentioned here, these are important factors to compose a successful business model. However, collective intelligence, or simply put 'platform', is a must for the provision of O2O service. Therefore, it alone can be sufficient to explain the innovation of a business model. For example, the representative O2O services such as e-commerce and SNS can't exist without a platform. That is, it is less important for a new and innovative business model than disruptive technology, long tail, and personalization.

Last, "Less is More" has the lowest weight in the analysis. Less is more can be applied to some business models but can't to others. The best examples are Apple's iOS and Google's Android, which are the two axes of smartphone operation systems. The core value of iOS-based services is simplicity because of its closed policy of platform operation. Meanwhile, Android-based operation follows open platform, inviting several manufacturers and developers in it, which has resulted in the features of Google: diversity and complicity. Some customers are for iOS and others prefer Android. It is the matter of a customer's preference. Therefore, the 'less is more' strategy cannot be a cure-all in the market. This is why this factor ranks in the lowest in priority.

In summary, the following is needed to build an innovative business model in the era of the 4th Industrial Revolution: first, apply disruptive technology to make a creative business paradigm. Second, place weight ion long tail and personalization and offer customers customer-specific optimal products and services. Last, it is important to establish a platform utilizing 'Hyper Connectivity'. This series of concepts can be a direction to a successful business model for startups when they develop or launch a new business (products and services) in the. To survive the heated competition in a market, it is very important for startups to know which value should be offered first. This study offers them the areas and perspectives to consider when facing the key decision.

In this study, e-business domain for research wasn't limited to a certain sector. Instead, the overall areas of e-business including O2O service were covered. It is because there are not many precedent studies for an innovative business model and thus, it was felt necessary to do an exploratory study first on the overall areas of e-business. Therefore, the following study is expected to divide or narrow down business area to study and find new factors to consider for a successful business model. In addition, it will be better to carry out correlation analysis between the factors.

References

1. CB INSIGHTS, https://www.cbinsights.com (2017)
2. Minwha, L., Aeseon, K.: Technology model for the 4th industrial revolution, AI + 12 Tech. Korean Inst. Commun. Inf. Sci., Inf. Commun. **34**(8), 3–8 (2017)
3. Aeseon, K., Minwha, L.: Understanding on the 4th industrial revolution and innovation strategy of enterprises from the viewpoint of human desire. In: Integrated Conference Treatise Compilation by the Korean Academic Society of Business Administration, pp. 1363–1376 (2017)
4. Govindarajan, V., Govindarajan, T., Stepinski, A.: Why unicorns are struggling. Harvard Bus. Rev. (2016)
5. Santos, J., Spector, B., Heyden, L.V.D.: Toward a theory of business model innovation with incumbent firms. Working paper no. 2009/16/ EFE/ST/TOM, INSEAD, Fontainebleau, France (2009)
6. Zott, C., Amit, R.: The fit between product market strategy and business model: implications for firm performance. Strateg. Manag. J. **29**(1), 1–26 (2008)
7. Zott, C., Amit, R.: Business model design: an activity system perspective. Long Range Plan. **43**(2–3), 216–226 (2010)
8. Margretta, J.: Why business models matter. Harvard Bus. Rev. **80**(5), 86–92 (2002)
9. Hienerth, C., Keinz, P., Lettl, C.: Exploring the nature and implementation process of user-centric business models. Long Range Plan. **44**(5–6), 344–374 (2011)
10. Vanhaverbeke, W.: The interorganizational context of open innovation. In: Chesbrough, H., Vanhaverbeke, W., West, J. (eds.) Open Innovation: Researching a New Paradigm. Oxford University Press, Oxford (2006)
11. Foss, N.J., Laursen, K., Pedersen, T.: Linking customer interaction and innovation: the mediating role of new organizational practices. Organ. Sci. **22**(4), 980–999 (2011)
12. Salge, T.O., Bohne, T.M., Farchi, T., Piening, E.P.: Harnessing the value of open innovation: the moderating role of innovation management. Int. J. Innov. Manag. **16**(03), 1–26 (2012)
13. Keinz, P., Hienerth, C., Lettl, C.: Designing the organization for user-driven innovation. J. Organ. Des. **1**(3), 20–36 (2012)
14. Jansen, J., Bosch, V.D., Volberda, H.W.: Managing potential and realized absorptive capacity: how do organizational antecedents matter? Acad. Manag. J. **48**(6), 999–1015 (2005)
15. Bengtsson, M., Kock, S.: Co-opetition in business networks: to cooperate and compete simultaneously. Ind. Mark. Manag. **29**, 411–426 (2000)
16. Brandenburger, A.M., Nalebuff, B.J.: Co-opetition. Currency/Doubleday, New York (1996)
17. Daft, R.L., Lewin, A.Y.: Where are the theories for the "new" organizational forms? An editorial essay, Organ. Sci. **4**(4), i–vi (1993)
18. Dunbar, R.L.M., Starbuck, W.H.: Learning to design organizations and learning from designing them. Organ. Sci. **17**, 171–178 (2006)
19. Mendelson, H.: Organizational architecture and success in the information technology industry. Manag. Sci. **46**, 513–529 (2000)
20. Brynjolfsson, E., Hitt, L.: Intangible assets and the economic impact of computers (2004)
21. Morris, M., Schindehutte, M., Allen, J.: The entrepreneur's business model: toward a unified perspective. J. Bus. Res. **58**(6), 726–735 (2005)
22. Chesbrough, H.: Business model innovation: it's not just about technology anymore. Strategy Leadersh. **35**(6), 12–17 (2007)

23. Zott, C., Amit, R., Massa, L.: The business model: recent developments and future research. J. Manag. **37**(4), 1019–1042 (2011)
24. Teece, D.: Business models, business strategy and innovation. Long Range Plan. **43**(2–3), 172–194 (2010)
25. Richardson, J.: The business model: an integrative framework for strategy execution. Strateg. Chang. **17**(5–6), 133–144 (2008)
26. Osterwalder, A., Pigneur, Y.: Designing business models and similar strategic objects: the contribution of IS. J. Assoc. Inf. Syst. **14**(5), 237 (2013)
27. Rayna, T., Striukova, L.: From rapid prototyping to home fabrication: how 3D printing is changing business model innovation. Technol. Forecast. Soc. Chang. **102**, 214–224 (2016)
28. Casadesus-Masanell, R., Ricart, J.E.: From strategy to business models and ontotactics. Long Range Plan. **43**(2), 195–215 (2010)
29. Chesbrough, H.: Business model innovation: opportunities and barriers. Long Range Plan. **43**(2–3), 354–363 (2010)
30. Zott, C., Amit, R.: Measuring the performance implications of business model design: evidence from emerging growth public firms. Working paper 2002/13/ENT/SM, INSEAD, Fontainebleau, France (2002)
31. Holm, A.B., Gunzel, F., Ulhoi, J.P.: Openness in innovation and business models: lessons from the newspaper industry. Int. J. Technol. Manag. **61**(3), 324–348 (2013)
32. Abdelkafi, N., Makhotin, S., Posselt, T.: Business model innovations for electric mobility: what can be learned from existing business model patterns? Int. J. Innov. Manag. **17**(01) (2013)
33. Drucker, P.F.: The Practice of Management. Harper Brothers, New York (1954)
34. Marcau, L.M.: Human rights challenges posed by the fourth industrial revolution. The Uber Case. MS thesis (2017)
35. Gorevayaa, E., Khayrullinaa, M.: Evolution of business models: past and present trends. Procedia Econ. Finan. **27**, 344–350 (2015)
36. Kroik, J., Skonieczny, J.: The use of business models in forming corporate social responsibility. Research Papers of Wrocław University of Economics (2015)
37. Pourdenhad, J., Baker, D.: Crowdsourcing Business Model Innovation. Using Social Media Platforms and Design (Work in progress) (2011)
38. Howe, J.: The rise of crowdsourcing. Wired **14**(6) (2006a)
39. Howe, J.: Crowdsourcing: a definition blog crowdsourcing. http://crowdsourcing.typepad.com/cs/2006/06/crowdsourcing_a.html (2006b)
40. Neto, C., Ana, M., Ana, E.T.: Emerging collective intelligence business models. In: MCIS Proceedings (2012)
41. Pilloni, V.: How data will transform industrial processes: crowdsensing, crowdsourcing and big data as pillars of industry 4.0. Future Internet **10**(24) (2018)
42. Sandulli, F.D., Chesbrough, H.W.: The two faces of open business models. SSRN Working Paper, 1–25 (2009)
43. West, J., Gallagher, S.: Challenges of open innovation: the paradox of firm investment in open-source software. R&D Manag. **36**(3), 319–331 (2006)
44. Chesbrough, H.W.: Open Innovation: The New Imperative for Creating and Profiting from Technology. Harvard Business Press, Boston, MA (2003)
45. Chesbrough, H.W.: Open Business Models: How to Thrive in the New Innovation Landscape. Harvard Business Press, Cambridge, MA (2013)
46. Zimmermann, S.: Industry 4.0 and open innovation. ATOS blog. https://atos.net/en/blog/industry-4-0-open-innovation (2018)
47. Howe, J.: Crowdsourcing: Why the Power of the Crowd Is Driving the Future of Business. Crown Business, New York, NY (2008)
48. Daren, B.C.: Crowdsourcing as a model for problem solving. Convergence **14**, 75–90 (2008)
49. Kavadia, S., Ladas, K., Loch, C.: The transformative business model: how to tell if you have one. Harvard Bus. Rev. (2016)
50. Afuah, A.: Business model innovation: concepts, analysis, and cases. Routledge (2014)

51. Schendel, D.: Introduction to the special issue on corporate entrepreneurship. Strateg. Manag. J. **11**(Summer Special Issue), 1–3 (1990)
52. Lendel, V., Varmus, M.: Creation and implementation of the innovation strategy in the enterprise. Econ. Manag. **16**, 819–825 (2011)
53. Kroik, J., Skonieczny, J.: The use of business models in forming corporate social responsibility. Research Papers, Wrocław University of Economics (2015)
54. Matthew E.M.: The less-is-best approach to innovation. Harvard Bus. Rev. (2012)
55. Walsh, S., Linton, J.: Infrastructure for emerging markets based on discontinuous innovations. Eng. Manag. J. **12**(2), 23–31 (2000)
56. Kostoff, R.N., Boylan, R., Simons, G.R.: Disruptive technology roadmaps. Technol. Forecast. Soc. Chang. **71**, 141–159 (2004)
57. Woo, C.S., Gim, G.Y., Kang, S.P.: Comparative study on bankruptcy prediction model using LOGIT and AHP analyses. Korean J. Finance Manag. **7**(2), 229–252 (1997)
58. Bin, J.Y.: ANP model for analyzing the priority of risk factors in water supply networks. M.D. thesis of Korea University (2005)
59. Park, H., Goh, G.G., Song, J.Y., Shin, G.S.: A Study on Applying Multi-criteria Analysis to Prefeasibility Survey. Public Investment Management Center of Korea Development Institute, Seoul (2000)
60. Zahedi, F.: The analytic hierarchy process—a survey of the method and its applications. Interfaces **16**(4), 96–108 (1986)
61. Saaty, T.L.: The Analytic Hierarchy Process. Mcgraw Hill, New York (1980)
62. Saaty, T.L.: Decision Making with Dependence and Feedback: The Analytic Network Process. RWS Publications, Pittsburgh, PA (1996)
63. Gim, G.Y.: AHP use in project management. J. Korean Inst. Project Manag. Technol. **7**(2) (1997)

A Study on the Efficiency of Global Major Mobile Operators

Jeongil Choi, Youngju Park and Yonghee Kim

Abstract This study analyzed the efficiency of global major mobile operators and the reason for their efficiency. For this purpose, the financial data of 96 operators in 40 OECD member countries were utilized. Based on this financial data, this study conducted a comparative analysis of the efficiency among operators and among countries. As for the analysis method, productivity was estimated, using the Malmquist Index. Operation efficiency and technology efficiency were estimated to analyze what elements acted on the productivity. The analysis results revealed that the productivity of mobile operators was led by technology efficiency. In verification among the countries, there was also cause for productivity enhancement; this was due to the process of conversion from 3G to 4G or M&A, but effective competition should be formed to increase productivity.

Keywords Efficiency · Malmquist productivity · Mobile telecommunications Technical changes

1 Introduction

Recently, the fourth industrial revolution has become a conversation topic in most industrial circles. Discussion is particularly noticeable in the telecommunications industry, a combination of wireless and wired communications.

The telecommunications industry, based on the development of the network speed represented by 5G, has considerable influence and has created a ripple effect on

J. Choi · Y. Kim (✉)
College of Business Administration, Soongsil University, Seoul, South Korea
e-mail: yh.kim@soongsil.ac.kr

J. Choi
e-mail: jichoi@ssu.ac.kr

Y. Park
Graduate School of Business, Soongsil University, Seoul, South Korea
e-mail: herver@ssu.ac.kr

R. Lee (ed.), *Software Engineering Research, Management and Applications*, Studies in Computational Intelligence 789, https://doi.org/10.1007/978-3-319-98881-8_9

all industries. It is expected that telecommunications would be the basis that leads innovations in other industries. Just as the conversion from 2G to 3G, and from 3G to 4G became faster, it is expected that the conversion from 4G to 5G will be faster yet. Based on this, 5G-based technologies, called the fourth industrial revolution, will gradually lead to other developments, including artificial intelligence, autonomous driving, smart factories, etc.

The emergence of 5G with the development of network technology does not simply represent an increase in data transmission speed or in processing capacity. 5G can be said to be the infrastructure that creates completely new businesses and that brings about innovative changes. This is why the technological innovation of mobile operators, including the innovation of 5G, is accepted as having a destructive power that can completely change the appearance of our society.

Faster mobile communication speed improves the ability to process data faster in real time and enables the processing of more data. Since this can provide an environment that can build up various platforms, it is expected that present social media enterprises or portal business operators would compete for a new type of platform in the name of the fourth industrial revolution.

In the meantime, unlike such positive changes of technologies, in terms of operation, global mobile operators face a dilemma. This is because of a weakening growth dynamic resulting from telecommunication market saturation and because of the rise of powerful competitors from non-telecommunication businesses such as internet portal companies including Apple and Google. According to research by Ernst & Young [1], regarding a question about the quests faced by global telecommunication services, the replies of 73% indicated destructive competition and 64% indicated unclear regulation environments. Also, 93% of respondents anticipated that OTT businesses will change the demand pattern of the future telecommunication market.

In particular it is estimated that cable market sales of super high-speed internet will gradually decrease, and that sales in the wireless market will gradually increase. Through the release of new plans to maximize sales in a changing environment–such as plans with a main focus on data–global mobile carriers are actively advancing various combination businesses that expedite the use of data. Even so, on the other hand, they are plotting an increase in business efficiency through methods such as organization of non-profit businesses and the progression of business diversification to secure an economy of scale. Also, businesses are pouring a lot of energy into the creation of new services. Regarding infrastructure, they are working on the provision of frequency circuits and networks such as 4G and 5G, and regarding service they are providing portfolios such as IPTV, cloud for businesses, and smart homes. They are quickly changing the profit composition of mobile carriers.

Situations like these are not so different in South Korea. Due to an intensification of the cable and wireless replacement phenomenon, the cable telephone service has entered a path of decline. Even the mobile voice telecommunication service has stagnated in growth with the number of subscribers exceeding the number of citizens in 2013. Because of this, competition among mobile carriers has become progressively more intense. To survive in a heated, competitive market, carriers are making an effort to maximize collection profit and to create efficiency between the different parts of their business by expanding the application limits of the strengths (the main values) the company possesses. Increasing the business efficiency in an effort to find new motive power for such growth is the main business survival strategy of companies in a saturated subscriber-procuring market. Domestic mobile carriers are actually expanding their business domains to IoT service on one hand, Smart Home and broadcasting media fields, they are actively advancing a structure modification.

In 2014 KT went forth with a 28.3% large-scale structure modification of permanent employees and disposed of its unrelated business section, KT Rental. In 2015 SKT went forth with a 10% permanent employee modification and is making an effort to increase company efficiency by actions such as organizing unprofitable overseas businesses. Even so, SKT merged with SK Broadband as a 100% subsidiary company; although they failed in plans to take over CJ Hello Vision, they are continuously making an effort to diversify related businesses.

In the meantime, the share of the mobile communication market in South Korea is fixed and hardly changing; three companies concentrate on competing for the promotion to attract subscribers, rather than for service. This has caused a great deadweight loss.

This means that a market failure occurs due to the limitation in competition. It also means that a phenomenon develops in which the benefit occurring in the market does not come to economic subjects, other than to the three telecommunication companies, due to exaggerated advertisements or promotion costs.

Until now, the results in mobile communication business have been discussed mainly in terms of visible scales, such as the scale of revenue or the number of subscribers. Of course, it is easy to achieve a result through an economy of scale if there is a lot of revenue or there are many subscribers; however, the evaluation and research of the activity of distributing limited resources efficiently are as important as the results of the scale of revenue revealed externally.

Nevertheless, there is a slight interest in the efficiency of mobile communication companies and in the factors of determination within domestic academic circles. Also, in the past a few studies conducted in the field of economics focused on the high-speed Internet business or did not reflect changes in the mobile market well. Within media communication studies, there was little interest in the efficiency of the media industry related to mobile communication. Studies that were conducted analyzed superficial results in the market, such as the number of subscribers or the amount of revenue, rather than efficiency.

This study analyzes the efficiency of the world's major mobile carriers. Through an international comparison, it hopes to find measures for increasing the efficiency of South Korea's mobile carriers. To do this, through the Malmquist methodology, the productivity between 2012 and 2016 was calculated. The objective of this paper is to analyze the relative change in productivity of mobile carriers and to review a path for the enhancement of productivity of South Korea's mobile carrier industry.

For this purpose, research questions were set as follows.
RQ 1: How is the management efficiency of the domestic mobile carriers as compared to that of the world's major mobile carriers?

RQ 2: What are the factors determining the management efficiency of the world's major mobile carriers and the domestic mobile carriers and what are the differences between them?

2 Theoretical Background

2.1 Concept of Efficiency

In general, If the competition of enterprises in the market is limited, efficiency declines, and in a competitive market structure–in the long term–more efficient enterprises survive. Here, efficiency refers to the ratio of inputs and outputs. In other words, how effectively have the inputs consumed been used and combined in the process of producing outputs, corresponding to the minimum unit costs, which is divided into cost efficiency and scale efficiency.

Cost efficiency means whether a specific output was produced by the combination of inputs composed of the lowest prices. Cost efficiency is divided into technical efficiency, which shows whether the production was made on the most efficient producible curve, and allocative efficiency, which shows whether the production was made by allocating inputs at the minimum unit costs. In contrast, scale efficiency refers to whether the production scale is the optimal scale when outputs are produced at the minimum unit costs. At this time, if benefits of scale decrease or increase, scale inefficiency occurs. If the benefits of scale increase, economies of scale occur; if the benefits of scale decrease, diseconomies of scale occur.

2.2 Productivity Estimation Method

The Data Envelopment Analysis (DEA) calculation method is a nonparametric calculation method that has the advantages of flexibility and wide use because it does not require a particular form of function or the ability to consider several inputs and outputs compared to other productivity calculation methods. In the DEA calculation, first in relation to the efficiency calculation, there is the method of calculating the isotropic curve by DEA using the actual output and input combination and then using the efficiency index calculation method to measure the amount of efficiency by calculating the distance between productivity points inside and the productivity points of the isotropic curve. Also, using DEA, there is the Malmquist productivity variation index calculation method is used to evaluate the productivity improvement by measuring the degree of change in comparison to the past year.

A review of the literature indicated studies that used the nonparametric calculation method to calculate the change in productivity in manufacturing and finance. In the area of energy there has been an analysis of the level of productivity in the oil and gas industry. Literature reviewed regarding the productivity analysis of the area of power was focused either on the dominating points of past vertical integration structure or on an analysis of only particular years due to the lack of materials.

The Malmquist Productivity Index, based on the output distance function proposed by Shepherd [2] to measure the efficiency, was defined by Caves et al. [3], and a method for measurement using a linear planning model was developed and utilized by Färe et al. [4]. The Malmquist Productivity Index model is a transfiguration of DEA with the foundation of distance function and calculates the efficiency change of the DMU between two different points through a longitudinal and cross-section analysis.

The Malmquist Productivity Index can be calculated under an assumption of Variable Returns to Scale (VRS) and Constant Returns to Scale (CRS) under an input-oriented and output-oriented model. The productivity index with a basis on the skill level between t and t + 1 is the following [3]: (xt, yt) and (xt + 1, yt + 1) shows the combination of input (x) and output (y) of t and t + 1. The distance function of t and the distance function of t + 1 each shows the distance on the combination of input (x) and output (y) from the skill level (efficient frontier) of each t and t + 1.

$$M^t = \frac{D_c^t\left(x^{t+1}, y^{t+1}\right)}{D_c^t(x^t, y^t)} \quad (1)$$

$$M^{t+1} = \frac{D_c^{t+1}\left(x^{t+1}, y^{t+1}\right)}{D_c^{t+1}(x^t, y^t)} \quad (2)$$

Färe et al. [4] defined the input standard for the Malmquist productivity variation index as follows.

$$\begin{aligned} & M\left(x^{t+1}, y^{t+1}, x^t, y^t\right) \\ & = \left[\frac{D_c^t\left(x^t, y^t\right)}{D_c^t\left(x^{t+1}, y^{t+1}\right)} \times \frac{D_c^{t+1}\left(x^t, y^t\right)}{D_c^{t+1}\left(x^{t+1}, y^{t+1}\right)}\right]^{\frac{1}{2}} \\ & = \left[\frac{D_c^{t+1}\left(x^t, y^t\right)}{D_c^{t+1}\left(x^{t+1}, y^{t+1}\right)}\right] \\ & \times \left[\frac{D_c^{t+1}\left(x^{t+1}, y^{t+1}\right)}{D_c^t\left(x^{t+1}, y^{t+1}\right)} \times \frac{D_c^{t+1}\left(x^t, y^t\right)}{D_c^t\left(x^t, y^t\right)}\right]^{\frac{1}{2}} \end{aligned} \quad (3)$$

The first equation above shows the productivity variation between t and t + 1 and is interpreted as the total factor productivity. This means that indices greater than 1 indicates an increase in productivity; if less than 1 it indicates a decrease in productivity; and if it is 0, then there is no variation in the productivity. The Malmquist Productivity Index can be sorted into the Technical Efficiency Change Index (TECI) and the Technical Change Index (TCI) under the proximation of CRS.

In the second equation above, the first term calculates the variation in the technical efficiency between t and t + 1 of the ratio of two distance functions and the second term shows the movement of manufacture technique, hence the variation in technique. The TECI is the result of the division of the technical efficiency of t + 1 and the technical efficiency of t; it shows the degree of productivity variation that contributes to the change in the efficiency variation between the two. The combination of input and output shows whether this is closer (TECI > 1) or further (TECI < 1) from the efficiency frontier. The TCI evaluates, then finds, the geometric mean of the technique variation between t and t + 1 and shows the degree of productivity variation contribution in the technique variation of the two.

It can be seen that if the number of indices is larger than one, the technique has advanced, if lower than one the technique has degenerated and the same as one the technique has stopped. TCI is shown by such exterior effects such as government policies, physical technique and economic environment change and the TECI shows increase arising from the efficient combination of interior aspects.

3 Data Collection and Analysis

The measurement of variables is very important in an analysis of efficiency. For the most ideal measurement, it is best to use all input elements and all output elements. However, due to realistic constraints or the degree of freedom, inevitably a model is set up that includes only several important input and output elements. At this time, since the value of efficiency may differ depending on the selection of variables and the unit, it is necessary to select safe and appropriate variables and to consider variables used in existing studies.

Studies using DEA in an analysis of efficiency of the communication industry and broadcasting industry include Zhu [5] and Sueyoshi [6]. Considering the convergence of the communication and broadcasting industries, this study would draw variables comprehensively, taking into account studies of the efficiency of the communication industry and that of the broadcasting industry.

Sueyoshi [6] analyzed the efficiency of the communication industry of OECD-member countries with revenue as an output and with the number of communications networks, the amount of investments in the field of communication, and the number of employees in the field of communication as inputs. There is a study that used the total assets, instead of the amount of investments, as an input. The domestic studies mainly put revenue as an output and put the number of communications networks, the number of employees, and the total amount of investments as inputs; some used assets and operating costs as inputs.

Thus, this study, too, would use communication carriers' revenue, which was mainly used in the preceding studies, as the basic output. For input elements, for capital (K), variables such as intangible assets, depreciation, and financial expenses, were used, while for labor (L), employees were selected to analyze efficiency.

3.1 Data

As data needed for this study, GICS Code 5010-was selected based on the OSIRIS Database [7] that includes the financial statements of about 30,000 listed companies in 120 countries around the world. Data where the calculation in the OSIRIS was difficult, or data that differed from the definition of financial statements were supplemented by each company's annual report. This study was conducted by analyzing the present condition of 96 major mobile carriers.

Since it is very complicated to compare individual operators' productivity, this study aims to compare the productivity results of the mobile communication industry in each country based on the results of individual operators' productivity, rather than to compare individual operators' productivity therefore this study compared results by country, calculating the index of productivity of each country in each year (Table 1).

3.2 Estimation Result of Malmquist Productivity Change Index

The MPI is largely separated into two indexes, the first being the Efficiency Change index and the other being the TCI. The averages that this study analyzed during a 5-year period are shown in the following Table 2.

As a result of the analysis, the efficiency of three major countries and South Korea's efficiency are analyzed as follows:

India's efficiency was the highest. As a result of an analysis through the average for five years, operation efficiency increased in India's mobile market through the rapid expansion of the market along with the introduction of 4G service. In India, demand for related services sharply increased with the first operator Airtel's launching of 4G service in 2016. Accordingly, India's mobile industry continues to experience an

Table 1 Descriptive statistics on inputs and outputs of the model[7]

Year	Input				Output
	Intangible fixed assets	Depreciation	Financial expenses	Staff	Revenue
2016 average	12,660,866	3,190,154	677,746	48,405	18,016,731
2015 average	12,439,134	3,123,632	632,200	47,933	17,953,441
2014 average	11,279,947	3,003,298	633,044	47,013	18,000,568
2013 average	11,115,313	3,229,173	623,055	46,735	18,657,198
2012 average	10,292,107	3,574,040	599,365	46,036	18,462,129
Total average	11,557,473	3,224,059	633,082	47,225	18,218,014

*Intangible fixed asset, Depreciation, Financial expense, Revenue: Thousand dollars

Table 2 The accumulated growth of productivity and its determinants between 2012 and 2016

Country	Malmquist	Effch	Tech	Scalech
IN	1.053	1.002	1.053	0.984
FR	1.048	0.995	1.059	0.997
GB	1.020	0.979	1.052	1.000
CN	1.004	0.964	1.044	0.969
HK	1.006	0.963	1.049	0.987
ID	0.973	0.959	1.014	0.967
GR	1.031	0.946	1.090	1.002
US	1.009	0.946	1.070	0.953
SE	1.016	0.944	1.073	0.976
JP	1.021	0.943	1.086	0.938
TH	1.015	0.939	1.079	0.914
RU	1.024	0.935	1.096	1.023
HU	1.003	0.932	1.075	0.987
QA	0.999	0.930	1.077	1.310
PL	1.007	0.928	1.085	1.078
TR	0.997	0.927	1.078	1.013
AU	1.010	0.924	1.094	0.925
AE	0.990	0.921	1.074	0.893
DE	0.963	0.920	1.015	0.909
AT	0.968	0.919	1.056	0.969
CA	0.987	0.916	1.079	0.950
MA	0.999	0.916	1.091	0.919
EG	0.985	0.910	1.081	0.972
KR	0.999	0.908	1.099	0.911
MY	0.951	0.908	1.047	0.967
TW	0.964	0.905	1.067	0.984
PE	0.983	0.901	1.091	0.977
ZA	0.989	0.900	1.099	0.948
ES	0.952	0.896	1.061	0.924
SG	0.975	0.894	1.093	0.948
AR	0.979	0.893	1.097	0.995
CH	0.966	0.892	1.083	0.904
NZ	0.955	0.892	1.075	1.025
NL	0.967	0.889	1.087	0.904
KW	0.937	0.889	1.060	0.912
MX	0.973	0.888	1.094	0.887
BE	0.942	0.886	1.065	0.933
PH	0.923	0.883	1.044	0.916
CL	0.958	0.883	1.086	0.938
NO	0.946	0.881	1.075	0.969
IL	0.961	0.880	1.094	0.938
DK	0.914	0.875	1.049	0.896
IT	0.890	0.869	1.030	0.881
BR	0.893	0.840	1.066	0.943
SA	0.904	0.835	1.080	0.858
Total	0.992	0.929	1.069	0.961

explosive annual growth higher than 10%. In particular, with the increase of demand for telecommunication related services, the expansion and growth of the market and the mobile e-commerce utilizing mobile platform is accelerated. Digital payment is spreading, which has a great impact on the entire Indian society [8].

In addition, as the competition between telecommunication companies became fierce, since September 2016, Reliance Communication has done aggressive marketing, launching the 4G service 'Jio' and quickly increased its market share, providing unlimited free calls. Through this, it has currently secured 130 million subscribers.

In India's telecommunication market, as demand for various services based on wireless data explosively increased since the launching of 4G, demands for smartphones and telecommunication equipment have rapidly increased, and the growing tendency of operation efficiency and technology efficiency based on this is remarkable. In addition, this growth leads to the expansion of the market of 4G-based applications. Along with this, India's major mobile communication companies' reorganization of the market is progressing through M&A, so it is expected that operation efficiency will increase further, and considering the telecommunication market, which is still incomplete as compared to the population size, it is judged that the efficiency of the Indian mobile operators will continue to increase.

In the next case of France, since Free Mobile's entry in 2012, an enormous change began in terms of share, result, and stock prices. The success of the fourth telecommunication policy towed policies such as the mandatory roaming of the existing operators and the markdown of connection charges.

Free Mobile in France participated in the mobile market, oligopolized by three companies, including France Telecom's 'Orange,' 'SFR' of which the share was divided into the world's top 5 complex media groups, Vivendi Group and the global telecommunication company, Vodafone, and a subsidiary of Bouygues SA, 'Bouygues Telecom' in earnest as the fourth mobile operators in January 2012 and has now jumped up to be the operator ranked third. Free Mobile could succeed because its mother company was the second operator of high-speed Internet in France. It had the experience, infrastructure and awareness of the telecommunication business and could provide mobile services inexpensively as a combined product, which became the biggest driving force [9].

At the time, Free Mobile created a sensation with a flat rate product at 19.99 euros for unlimited use of voice calls and data through the sophistication of the network so that it could attract 60% of new subscribers. The factor of Free Mobile's success was that it was able to provide the service at a rate relatively lower than that of its competitors by combining a mobile device with high-speed Internet in spite of above average rates. It is judged that as mentioned above, the company could lower the price through network sophistication (the increase of technology efficiency). Through this, effective competition and a low Herfindahl-Hirschman Index (HHI) appeared, so a strategic factor was generated for promoting the development of technology and the increase of operation efficiency, which resulted in a high productivity rank.

Next, in the U.K., according to Ofcom, the number of 4G subscribers increased continuously until the second quarter of 2017, and the 4G subscribers amounted to almost two thirds of all subscribers. In addition, with the conversion from 3G to

4G, the technology efficiency increased, so revenue has also steadily increased. It is judged that the productivity improves through this. It is operated in the way of which the telecommunications networks of British Telecom, which was formerly a national-run telecommunication operator, are leased to multiple telecommunication operators, including EE, O2, and Vodafone, etc., and recently, as BT took over EE, Hutchison would take over O2, but it failed due to the opposition of the European Commission being concerned about the decrease of the number of MNO operators into three.

Thus, the result in the same context can be inferred from the cases of France and South Korea. In South Korea, as the number of telecommunication operators decreased from five to three, operation efficiency also fell. In contrast, in France, the number increased from three to four, which brought about the decline of market concentration, so effective competition revived, which resulted in the increase of technology efficiency and productivity through the spread of technology investment. Mobile telecommunication authorities in the U.K. and Europe, also opposed to M&A, worried about the occurrence of restriction on competition based on this concern, and through this, effective competition continued and productivity could increase thanks to the technological revitalization through technology competition. If Hutchison's O2 merger was allowed, consequently, the variety of choice might decrease for consumers in the U.K. and the service quality could also become less. Thus, it is judged that mobile telecommunication rates probably increased through a combination of factors, which would very likely have acted as an element disturbing the increase of productivity.

Figure 1 examines the change of efficiency in major countries with developed mobile communications. It is a graph that supports the above results, and it is judged that the 2014–2015 efficiency was high in Japan because the operation efficiency increased as Softbank took over Sprint in 2013–2014, but a more in-depth study is necessary. The 2015–2016 efficiency increased in South Korea because the increase of productivity was reflected with the migration from 3G to 4G. As a result of a comprehensive analysis of studies on average for five years, it is found that the results followed stable technology advancement. In addition, it turned out that in all leading countries for productivity such as France, China or Hong Kong, technology advancement was higher than operation efficiency, which led to the improvement of productivity.

It is worth noting that although South Korea, the Republic of South Africa and Argentina had a high technical advancement, the operation efficiency was comparatively low and as a result failed in productivity improvement. This is especially true in the case of South Korea because although there was a high level of technical advancement, it showed the lowest level of operation efficiency and as a result demonstrated that the productivity improvement has stopped (productivity: 0.999).

Although South Korea's speed of technical advancement from the 2G market to the 4G LTE market has proved global, a large deadweight loss is being produced due to a great marketing cost consumption in comparison to the business size in order to acquire an adequate market share. This results in a competition limit which can be seen as the process of market failure. Furthermore, due to excessive advertisement

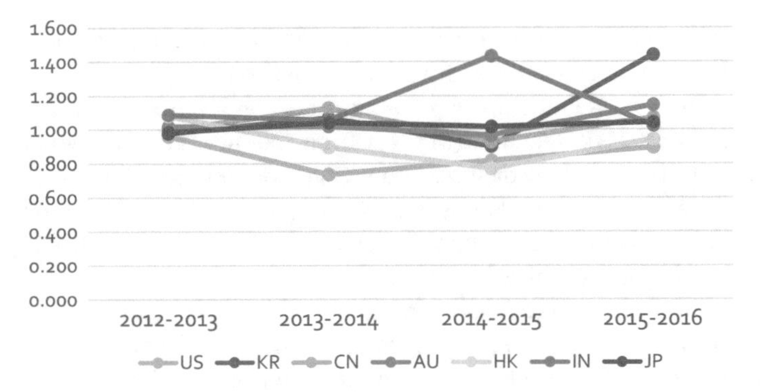

Fig. 1 Productivity growth in major telecom countries

or promotional costs, it can be seen that the situation where none of the economic subjects are benefiting is steadily expanding.

4 Conclusion

Despite there being a continuous improvement effort of the mobile telecommunication industry productivity level from the middle to late 2000s, it is understood that because of excessive competition inside the market, from 2012 a low operation efficiency is shown in comparison to comparatively high technical advancement compared to overseas mobile carriers of major countries. According to the MPI calculation results, the mobile telecommunication industry productivity level, rather than due to an increase in efficiency, has advanced due to a continuous effort in technical advancement effort.

Because there is a lot of room in the future for the occurrence of inefficiency following companies' operation or operation aspects, competitive market and the variation form of technical advancement, continuous effort for productivity improvement is needed.

As a result of summing up the results of this study, a conclusion was reached that effective competition should increase to improve the productivity of the communication market in each country. In other words, the appropriate number of mobile operators is needed in the market, and it is judged that they should launch mobile communication products at reasonable prices, holding each other in check, so that technological investment for service sophistication can be revitalized.

Through this conclusion, suggestions for the revitalization of the South Korean mobile communication market can be made as follows:

The government's attempt to lower mobile charges through policy pressure is made because mobile communication service operators' market share has been maintained firmly in the 5:3:2 structure for a long time. Individual operators use a business model optimized for their own market share, rather than the expansion of share through competition, and they strive for policy-dependent business to maintain the market without a rate competition or service innovation.

To solve this, the South Korean government actively introduced and supported Mobile Virtual Network Operators (MVNO's), but they have not yet been evaluated as effective competitors. OECD member countries introduced regulations on wholesale as one of the options for regulations to solve the problem of operators' domination of the mobile market or the oligopoly structure. South Korea, too, has introduced MVNO's to this purpose since 2012 [10].

Consequently, it was noted that, with the increase of the number of operators in the mobile market, HHI representing industrial concentration more or less decreased, but revenue did not flow from the existing mobile operators into MVNO's to that extent. There is doubt whether the actual consumer satisfaction increases and that tows the market demand, due to the difference in the bargaining power between the operators in the submarket, MVNO's and the existing mobile operators and institutional constraints. Therefore, MVNO's cannot produce reasonable MVNO exclusive products since they reckon the method of profit sharing with the existing mobile careers as access charge with LTE service, which is the main force of the present mobile communication industry, and the MVNO operators' LTE rate system is subordinate to mobile operators' rate systems.

Furthermore, MVNO services are perceived as inferior goods by the users. In addition, the number of offline contact points handling MVNO's, agencies and retailers are absolutely lacking, so service satisfaction is overall low, and they are not chosen by consumers.

The ultimate method to overcome this limitation, revitalize pure functions and lower the market concentration is the competitive fourth mobile operator's entry into the market. South Korea has discussed the establishment of the fourth mobile operator since 2011. The first consideration for the introduction of the fourth mobile operator continuously failed due to high investment costs such as frequency allocation charges, starting for the revitalization of stagnating WiBro.

The fourth mobile communication does not need to enter the market with a frequency allocated from the beginning. Policy support until the operator is settled in the mobile market through roaming would be just enough. And yet, practice is needed for stable business operation, and the MVNO industry can play such a role.

If the system related to MVNO's sufficiently improves, they can enter the market through resale, and they may first recruit subscribers in the form of full MVNO, which does not exist in the market now. In the oligopoly market, which has already been fixed in the U.K. or France, there is an experience in which the market was changed to an environment of effective competition, the price of use declined, and the service quality increased, thanks to competitive new subscribers entering the market.

To do so, it is necessary to adjust the government's policy in the direction in which it intervenes in the transactions (wholesale transactions) between MVNO's and mobile operators and maximizes consumers' profits to some extent.

In addition, it is necessary to make an effort to lower the high-level wholesale charges maintained between mobile operators and MVNO's to a level at which they can actually compete. Along with this, to overcome the limitation in the composition of products, it is necessary to quickly introduce a system for the advance purchase of massive data (purchase in bulk). MVNO's should also not save on investment in themselves so that they have the equipment and system to develop and set their own rates.

As the comparative productivity level of South Korea's mobile telecommunication market that has been analyzed up until now has been concluded from materials from a company level, certain limits exist. To supplement this, if materials on each mobile telecommunication industries of each companies can be used in the future, more research to compare the original results from the attained results using the same analysis method is necessary. Also, it is judged that more research linking future structural reform accomplishments and the variation in productivity is required.

References

1. Ernst & Young.: Global telecommunications study: navigating the road to 2020. Working Paper (2015)
2. Shepherd, R.W.:Theory of cost and production functions. Princeton University Press (1970)
3. Caves, D.W., Christensen, L.R., Diewert, W.E.: The econo mic theory of index numbers an d the measurement of input, output, and productivity. Econometrica. **50**(6), 1393–1414 (1982)
4. Färe, R., Grosskopf, S., Lindgren, B., Roos, P.: Productivity developments in Swedish hospitals: a Malmquist output index approach. In Data Envelopment Analysis: theory, Methodology, and Applications. Springer, Dordrecht (1994)
5. Zhu, J.: Imprecise DEA via standard linear DEA models' a revisit to a Korean mobile telecommunication company. Operation Res. **52**(2), 323–329 (2004)
6. Sueyoshi. T.: Stochastic frontier production analysis: Measuring performance of public telecommunications in 24 OECD counties. Euro. J. Operational Res. **74**(3), 466–478 (1994)
7. Van Dijk., B: Osiris DB. http://osiris.bvdifor.com
8. Lee, H.B.: India's mobile market. news.kotra.or.kr/user/globalBbs/kotranews/4/globalBbsDataView.do?setIdx=243&dataIdx=126542
9. Kim, T.J.: The fourth industry scenario drawn by the cable industry. www.zdnet.co.kr/news/news_view.asp?artice_id=20180413151649
10. Kwon, O.S.: Suggestions for the activation of domestic mobile communication market. http://www.etnews.com/20171127000186

A Study on Effects of Supporting Born Global Startups Policy Affecting the Business

Jung-Ran An, Sung Taek Lee, Ju-Hyung Kim and Gwang-Yong Gim

Abstract Since the economic growth rate has been led by large corporations, there has been a great deal of interest in the development of the economy. The rise of global startups with their focus on global markets and customers and who develop products/services that are globally focused are a catalyst for a new economic revival. This research has also been focused on corporate finance and nonfinancial performance. As a result of the analysis, it was found that space and accelerating support with loan funding policy affects entrepreneurial orientation. It was also found that the space, entrepreneurial education, and consulting, and accelerating support affects global orientation. Additionally, both entrepreneurial orientation and global orientation have a profound impact on financial performance and non-financial performance of global startups. This study not only provides meaningful information on the implementation of policy research and the implementation of startup policy, but it also provides a framework for the study.

Keywords Global startup · Startup policy · Entrepreneurial orientation
Market orientation · Network capabilities

J.-R. An · G.-Y. Gim (✉)
Department of Business Administration, Soongsil University, 369 Sangdo-Ro, Dongjak-Gu, Seoul 06978, Republic of Korea
e-mail: gygim@ssu.ac.kr

J.-R. An
e-mail: weareinthelord@hotmail.com

S. T. Lee · J.-H. Kim
Department of IT Policy and Management, Soongsil University, Soongsil University, 369 Sangdo-Ro, Dongjak-Gu, Seoul 06978, Republic of Korea
e-mail: totona22@ssu.ac.kr

J.-H. Kim
e-mail: bmckorea@empas.com

R. Lee (ed.), *Software Engineering Research, Management and Applications*, Studies in Computational Intelligence 789, https://doi.org/10.1007/978-3-319-98881-8_10

1 Introduction

1.1 Background and Purpose of Research

In Korea, there are increasing number of companies entering overseas due to limitations of market size and various leadership, organization, industry, and services. According to the 'Plan for Promotion of Global Business Startup' by the Korea Business Institute in 2013, 40.7% of venture companies have entered overseas business. Among them, simple exporters accounted for the highest percentage, followed by direct advancement and simple exports. In the 'Start-up ecosystem' report issued by the Start-up Alliance in 2016, the start-up advanced to overseas was 21.4% and the companies obtained actual sales were 11.9%. In general, companies seeking to enter the overseas market face difficulties due to various reasons such as language and cultural barriers, market information and lack of capital.

This study examined how significantly the policy support of this global start-up target affects the entrepreneurial orientation and global orientation of the start-up. As a result, this study is going to suggest policy support measures that affect corporate performance after checking if start-up's entrepreneurial orientation and global orientation weigh influence on financial/non-financial company performance. As a result, we expect the government's future global start-up support policy to provide a custom made strategy to help companies directly, and to succeed in the global market based on their unique characteristics and position.

2 Theoretical Background

2.1 Overview of Born Global Startups

Born Global Startups is a new start-up company that enters the overseas market at the beginning of its founding. These global start-ups mean start-ups that utilize the resources and networks of overseas markets held by companies in various countries and regions to quickly enter overseas markets and generate high profits. As a result of the study of this global start-up, various theories about the effects of this global start-up on the performance of the company have started to be established [1]. This global start-up has also been stimulated by new variables, as the existing 'Uppsala Internationalization Process Model' limits the explanations of SMEs' globalization. In particular, research has been conducted on factors such as global market orientation, marketing, international entrepreneurial orientation, and global experience of the company affecting the establishment of this global start-up [2].

2.2 Government's Start-up Support Policy

In a study by Han and Lee [3] on the impact of start-up on short-term and long-term performance through the support of the Start-up Incubator Center, the core services such as space, facilities, equipment, experts, technology, finance, laws, and networking support, it has been concluded that the presence of physical facilities or equipment has a significant impact on job creation and service satisfaction. There's also a study viewing that the results of start-up may be different depending on the level of support service provided by the Start-up Incubator Center. In a study of Cho and Park [4], the level of support service provided by the start-up incubator center was measured by using the laboratory and equipment that the center has as an index, and the performance of start-up varies according to the service level, and it has a positive effect on the management performance of start-up as the service level gets increased. In order to enhance the performance of the incubated companies, specialist support for the development of products/services with high market favorable prices and customer preferences and support for various promotional events are one of the important start-up support. Therefore, the start-up companies should plan various event support policies such as development of product/service, establishment of various distribution channels, formation of investment network, and systematic marketing support.

Start-up training means the activity to lead reach success by training those who prepare start-up and initial founders the planning of start-up, procedures to prepare after start-up, and method of business [5]. A systematic learning to culture founder's start-up ability is required in order to accomplish start-up's successful result. In order to understand the process of entrepreneurship and business, and to prepare for the risks that may arise in the process, we can help them through training [6], and also inspire confidence of founder. Therefore, there is a study that entrepreneurship education affects entrepreneurs' entrepreneurship intention and firm performance [7].

The start-up consulting gives advice, solution for the problems to business plan or operation and management of initial start-up companies or those who prepare start-up [8]. The reason for including consulting in the field of expertise in the start-up support project is that it is possible to present an alternative to overcome the various difficulties that the start-up business may face. The start-up consulting has various influences on the start-up companies. A study of the impact of entrepreneurship consulting on the financial performance of a firm shows that it affects sales growth and activity [9].

Accelerators that started in developed countries are characterized by transparent selection, fast investment, man-to-man professional mentoring and coaching. Currently, there are about 2000 accelerator companies around the world, and countries such as USA, UK, and Israel are active at the center. In Korea, it is defined as "a person who takes the main business of selection, investment and professional child-care of the initial founder, etc. as a person who is registered with the Small and

Medium Business Administration", and it made the accelerator register at the Small and Medium Business Administration.

Generally, start-up funding is classified into equity capital and borrowed capital depending on the type of capital attribution. Equity capital is the capital that an owner has unrestricted use of, and borrowed capital is the capital owned by the creditor and that must be repaid over the life of the contract. If start-up is insufficient to fund resources, it will have a negative impact on corporate performance [10]. However, in the case of start-ups that are small in scale and lack expert support, transaction performance and creditworthiness are so weak that it is difficult to apply for loans in the financial sector, and there are some cases where a relatively high interest rate or collateral not. Therefore, in many studies, start-up funding is considered to be an important factor directly affecting the start-up success [11]. In order to facilitate start-up and financing of SMEs, the government is implementing funding policies. In the case of early start-ups, the financing of the first financial sector is not easy because the scale is small and resources are always lacking. Therefore, it often happens that you pay high financial costs or depend on private bonds. This high barrier of finance is considered to be one of the main reasons for weakening the competitiveness of start-up companies. Therefore, the government is also using financing funds to provide financial support to entrepreneurs through credit guarantee and guarantee support.

2.3 *Entrepreneurial Orientation and Global Orientation*

Entrepreneurial Orientation refers to the process of seeking and developing innovative services that can be differentiated from other companies in the marketplace [12]. It is also defined as the degree to which new market opportunities are constantly pursued and efforts are made to improve and expand the current business [13]. There is also a research that global orientation narrows the culture and psychological distance of overseas market areas such as the degree of education of executives, overseas experience, and this is defined as operational type as well as organization culture that lead behavior to create a global value by making company members arrange and move forward global goals [14].

2.4 *Corporate Performance*

The most frequently used performance indicators in the previous studies related to the financial performance of this global start-up are economic indicators such as export sales growth, export ratio to sales, and export profitability [15, 16]. Some are measured by the ratio of overseas profit to total profit, concentration of R&D [17]. Reuver and Fisher (1997) used a single variable derived by combining all these variables, including the number of employees and the geographical range measured the degree of internationalization by combining various variables according to the theories of Sullivan (1994, 1996) and Ramaswamy and Rowthorn [18]. Many studies are based on non-financial aptitude, which emphasizes the recognition of management and personnel as one of the non-financial achievements of this global start-up. Knight et al. [19] measured international market performance as satisfaction with product performance such as market share, sales growth rate and pre-tax profitability [20]. And, in a study that examined the effect of this global company on strategic orientation and international performance, they measured subjective satisfaction for international activities in aspect of sales scale, market share, profitability, knowledge development as an index of international performance in cognitive aspect [21]. This study, too, carried out by taking variables of subjective performance, which emphasized non-financial performance emphasizing recognition of management and person in charge.

3 Research Design

The factors affecting the corporate performance of this global start-up are varied depending on the research perspective. There are insufficient in explaining the performance of existing traditional firms, such as research that they are determined by technology, entrepreneurship, and culture within the global start-up, and that external factors such as network and policy support are important. So this study examines the effects of current business start-up support on entrepreneurial orientation and global orientation, which are key factors of this global start-up, and also examines whether entrepreneurial orientation and global orientation have a significant effect on company performance. This study, then, operates in-depth analysis on how network competencies have a mediating effect when entrepreneurial orientation and global orientation affect corporate performance. The specific model to be considered in this study is shown in Fig. 1.

- Hypothesis 1-1: Space support among facility support will have a positive (+) effect on entrepreneurial orientation.
- Hypothesis 1-2: Event support among facility support will have a positive (+) effect on entrepreneurial orientation.
- Hypothesis 1-3: Supporting start-up training among professional support will have a positive (+) effect on entrepreneurial orientation.
- Hypothesis 1-4: Consulting support among professional support will have a positive (+) effect on entrepreneurial orientation.
- Hypothesis 1-5: Space support among facility support will have a positive (+) effect on entrepreneurial orientation.
- Hypothesis 1-6: Commercialization support among fund support will have a positive (+) effect on entrepreneurial orientation.
- Hypothesis 1-7: Policy loan fund support among fund support will have a positive (+) effect on entrepreneurial orientation.

- Hypothesis 2-1: Space support among facility support will have a positive (+) effect on entrepreneurial orientation.
- Hypothesis 2-2: Event support among facility support will have a positive (+) effect on entrepreneurial orientation.
- Hypothesis 2-3: Start-up training support among professional support will have a positive (+) effect on global orientation.
- Hypothesis 2-4: Consulting support among professional support will have a positive (+) effect on global orientation.
- Hypothesis 2-5: Accelerating support among professional support will have a positive (+) effect on global orientation.
- Hypothesis 2-6: Commercialization support among funding support will have a positive (+) effect on global orientation.
- Hypothesis 2-7: Policy loan fund support among funding support will have a positive (+) effect on global orientation.
- Hypothesis 3-1: Entrepreneurship orientation among company performance will positive (+) influence on the financial aspects of firm performance.
- Hypothesis 3-2: Entrepreneurship orientation among company performance will positive (+) influence on non-financial aspects of firm performance.
- Hypothesis 4-1: Global orientation among company performance will have positive (+) influence on the financial aspects of firm performance.
- Hypothesis 4-2: Global orientation will have positive (+) influence on the financial aspects of firm performance.
- Hypothesis 5-1: Network capability will control the relationship between entrepreneurial orientation and financial aspects of firm performance.
- Hypothesis 5-2: Network capability will control the relationship between entrepreneurial orientation and financial aspects of firm performance.
- Hypothesis 5-3: Network capability will control the relationship between global orientation and financial aspects of firm performance.
- Hypothesis 5-4: Network capability will control the relationship between entrepreneurial orientation and financial aspects of firm performance.
- Hypothesis 5-5: Network capability will control the relationship between entrepreneurial orientation and financial aspects of firm performance.
- Hypothesis 5-6: Policy support will control the relationship between entrepreneurial orientation and financial aspects of firm performance.
- Hypothesis 5-7: Policy support will control the relationship between global orientation and financial aspects of firm performance.
- Hypothesis 5-8: Policy support will control the relationship between global orientation and financial aspects of firm performance.
- Hypothesis 5-9: Project period will control the relationship between entrepreneurial orientation and financial aspects of firm performance.
- Hypothesis 5-10: Project period will control the relationship between entrepreneurial orientation and financial aspects of firm performance.
- Hypothesis 5-11: Project period will control the relationship between global orientation and financial aspects of firm performance.

- Hypothesis 5-12: Project period will control the relationship between global orientation and financial aspects of firm performance.

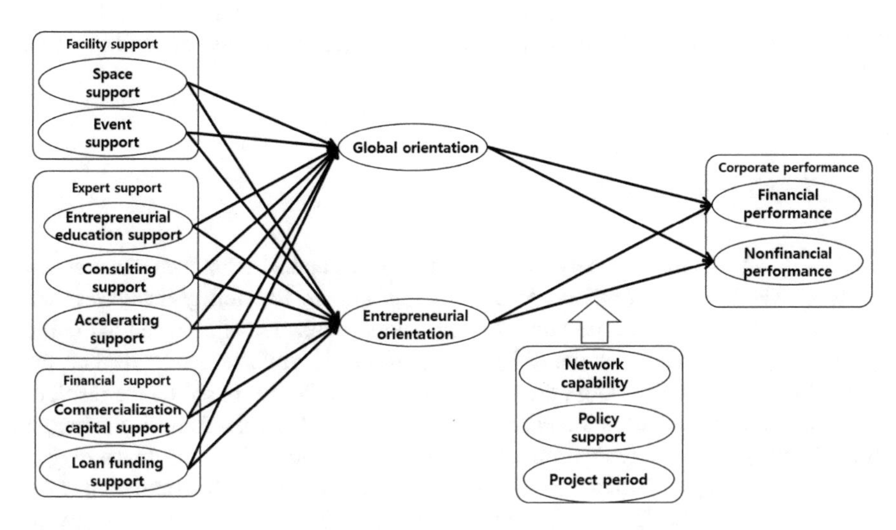

Fig. 1 Research model

4 Empirical Analysis

4.1 Data Collection and Analysis Methods

This study developed a research model based on the previous research and established research hypotheses to grasp the factors affecting the corporate performance of this global start-up policy support, and carried out an empirical analysis to verify each hypothesis. In order to collect data, questionnaires were used to measure the hypotheses and variables of the research model. A questionnaire survey was conducted targeting 230 officers and employees of global start-ups and potential global start-ups.

4.2 Characteristics of Samples

Frequency analysis was conducted to identify the demographic characteristics of the respondents. The number of analysis samples was 230, of which male was 61% and female was 29%. The most common age group was in their 30s (55%) and 230

(29%). The response rate was high with a 98% response rate in the 20–40s due to the tendency of start-up. Of the respondents, 62% said that they were office workers, followed by management position of 18%, and 77% said that they are doing global advancement, and 80% of the respondents said that they entered in period of less than 3 years after the start-up, showing appropriateness to the questionnaire for this global start-up.

4.3 Exploratory Factor Analysis

Reliability and feasibility analysis were conducted to confirm that the measurement tools for verifying the research model are appropriate. Reliability analysis is the process of confirming the possibility of achieving the same value when assuming repeated measurements for the object of the research, and this study verified the reliability of each item, using the cronbach's alpha coefficient based on the result of factor analysis. Cronbach's α means that the closer to 1 between 0 and 1, the higher the reliability. This study determined the threshold value at 0.7, and it was judged that there was no problem in the reliability of the variable when the coefficient value was 0.7 or more [22]. The measurement variable of this study was 0.7 or more, so it was confirmed that the internal consistency was secured. Validity is a measure of the extent to which a measurement tool has measured the concepts or attributes that it intends to measure originally and closely. In order to verify the feasibility, exploratory factor analysis (EFA) was used to investigate the degree of linking of inherent factors in observation variables. Factor loading of 0.5 or more was judged to be significant [23]. In this study, reliability analysis was performed to verify the internal validity of the extracted components, as a result of the analysis, Cronbach's Alpha value of all measured variables was 0.7 or more, securing reliability (Table 1).

4.4 Model Conformity Assessment

This study conducted a confirmatory factor analysis for the measured variables extracted through exploratory factor analysis using SMART-PLS 2.0. Unlike exploratory factor analysis, which is conducted for exploratory purposes, in order to grasp the direction of research in research that has not yet been systematized or established theoretically, confirmatory factor analysis is performed by setting the relationship between variables under the theoretical background and performing factor analysis [24]. When the composite reliability has a value of 0.70 or more and the mean variance extraction has a value of 0.5 or more, the reliability of the measured variables is obtained. A common method for measuring intensiveness is average variance extracted (AVE). AVE is the same as communality of construct. If AVE value is over 0.5 in average by using it the same as standard value used for individual indicators, it can be said to properly explain dispersion. The composite reliability

	Component							Cronbach's Alpha
	1	2	3	4	5	6	7	
Accelerating1	0.708							0.959
Accelerating2	0.692							
Accelerating6	0.678							
Accelerating3	0.676							
Accelerating7	0.643							
Accelerating4	0.608							0.948
Accelerating5	0.593							
Loan3		0.766						
Loan4		0.749						
Loan5		0.676						0.948
Loan1		0.583						
Loan2		0.582						
Event2			0.722					
Event4			0.708					0.923
Event3			0.656					
Event1			0.639					
Event6			0.625					
Event5			0.551					0.934
Space2				0.849				
Space3				0.792				
Space1				0.757				0.933
Space4				0.588				
Space5				0.572				
Training 3					0.669			
Training 1					0.651			0.92
Training 4					0.641			
Training 2					0.614			
Training 5					0.506			
Commercialization 1						0.666		0.933
Commercialization 3						0.558		
Commercialization 2						0.540		
Commercialization 5						0.469		
Consulting4							0.611	0.92
Consulting5							0.575	
Consulting2							0.490	
Consulting3							0.430	

Table 1 Reliability analysis and exploratory factor analysis

	요인				Cronbach's alpha
	1	2	3	4	
Non-financial 2	0.831				0.938
Non-financial 3	0.817				
Non-financial 1	0.792				
Non-financial 4	0.719				
Entrepreneurship3		0.805			0.888
Entrepreneurship4		0.750			
Entrepreneurship1		0.738			
Entrepreneurship2		0.702			
Entrepreneurship5		0.629			0.928
Global 2			0.845		
Global 3			0.773		
Global 1			0.739		
Global 4			0.637		
Financial1				0.835	0.857
Financial2				0.801	
Financial3				0.757	

Table 2 Composite reliability and average variance extracted

	Average variance	Composite reliability
Financial	0.689	0.867
Space	0.709	0.880
Event	0.750	0.938
Training	0.739	0.934
Consulting	0.742	0.920
Accelerating	0.777	0.946
Commercialization	0.778	0.933
Loan	0.787	0.937
Entrepreneurship	0.615	0.889
Global	0.766	0.929
Non-financial	0.795	0.939

(CR) and the average variance extracted (AVE) allow the reliability of the measured variables to be obtained when the composite reliability is greater than 0.70 and the mean variance extraction is greater than 0.5. [25] (Tables 2, 3, and 4).

Table 3 Model fit

	Standard		Result	
Redundancy	≧0(양수)		Global	0.188326
			Entrepreneurship	0.084391
			Non-financial	0.444141
			Financial	0.169605
Modelfit(R2)	0.26~	High	Global	0.396387
			Entrepreneurship	0.387874
	0.13–0.26	Middle	Non-financial	0.562164
	0.02–0.13	Low	Financial	0.294171
Total Model fit (R2 medium × Communality medium) Square root	0.36~	High	0.572350742	
	0.25–0.3	Middle		
	0.1–0.25	low		

4.5 Path Analysis

Hypothesis verification for the factors that this global startup policy support weighs on company performance was done through path coefficient for the research model (Table 5).

4.6 Hypothesis Verification

In order to verify each hypothesis of the research model, path analysis was conducted using the SMART-PLS 2.0 program. Analysis of the results was done by collecting the data, performing factor analysis using SPSS 2.0, and calculating the t value to detect path significance using model analysis and bootstrapping through PLS. This study adopted a significant hypothesis within the significance level of 5% (Table 6).

The results of the path analysis of the research model say that it dismisses all hypotheses that event support affects firms, orientation and global orientation. The hypothesis 1-3 that support for entrepreneurship education affects entrepreneurial orientation was dismissed. Hypothesis 1-6 that support for start-up training support affects entrepreneurial orientation and hypothesis 2-2 that event support affects global orientation were rejected. All other hypotheses were adopted.

Table 4 Reasonability of the identification of the identification factors

	Space	Education	Global	Entrepreneurship	Non-financial	Commercialization	Accelerating	Loan	Financial	Consulting
Training	0.706285									
Global	0.549522	0.537472								
Entrepreneurship	0.491176	0.523733	0.692485							
Non-financial	0.578781	0.52144	0.737389	0.608549						
Commercialization	0.668613	0.764919	0.553827	0.487427	0.530741					
Accelerating	0.719079	0.783888	0.562879	0.578105	0.557933	0.83268				
Loan	0.667456	0.81675	0.510354	0.526786	0.517442	0.831004	0.809713			
Financial	0.426026	0.423123	0.504039	0.493534	0.639803	0.514028	0.593626	0.475232		
Consulting	0.695917	0.831584	0.573763	0.582695	0.617516	0.808524	0.81967	0.807658	0.512241	
Event	0.771	0.801062	0.49878	0.479347	0.538016	0.778178	0.793339	0.771344	0.507536	0.799471

Table 5 The results of SMART-PLS 2.0-SEM

	Original sample (O)	Sample mean (M)	Standard deviation (STDEV)	Standard error (STERR)	T statistics (|O/STERR|)	P
Space support - > entrepreneur	0.179352	0.173459	0.084461	0.084461	2.123483	0.084
Space support - > global	0.165902	0.165733	0.074309	0.074309	2.232606	0.001
Event support - > entrepreneur	−0.031263	−0.026124	0.117503	0.117503	0.266064	0.206
Event support - > global	−0.069971	−0.067586	0.118932	0.118932	0.58833	0.058
Training support - > entrepreneur	0.0794	0.081197	0.116202	0.116202	0.683295	0.925
Training support - > global	0.154709	0.153968	0.088979	0.088979	1.738718	0.059
Consulting - > Entrepreneur	0.257177	0.255128	0.113285	0.113285	2.270183	0.005
Consulting - > Global	0.20058	0.204047	0.121521	0.121521	1.650575	0.044
Acceler - > entrepreneur	0.249265	0.243076	0.113793	0.113793	2.190513	0.006
Acceler - > global	0.202027	0.199331	0.112374	0.112374	1.797814	0.069
Commercialization - > Entrepreneur	−0.049297	−0.048875	0.127108	0.127108	0.387836	0.079
Commercialization - > Global	0.171012	0.157232	0.10026	0.10026	1.705681	0.092
Loan - > Entrepreneur	−0.086128	−0.075126	0.113969	0.113969	0.755715	0.073
Loan - > global	−0.185014	−0.171658	0.106431	0.106431	1.738339	0.225

(continued)

Table 5 (continued)

	Original sample (O)	Sample mean (M)	Standard deviation (STDEV)	Standard error (STERR)	T statistics (|O/STERR|)	P
Global - > financial	0.262929	0.258816	0.098286	0.098286	2.675139	***
Global - > non-financial	0.504943	0.513116	0.07857	0.07857	6.426634	***
Entrepreneur - > financial	0.267327	0.273249	0.100626	0.100626	2.656654	0.001
Entrepreneur - > non-financial	0.211696	0.199992	0.089736	0.089736	2.3591	0.011

*$p < 0.05$
**$p < 0.01$
***$p < 0.001$

Table 6 Causal relationship measurement by path analysis

	Original Sample (O)	Sample Mean (M)	Standard Deviation (STDEV)	Standard Error (STERR)	T Statistics (|O/STERR|)	P	Results
Space support - > global	0.262231	0.261364	0.080264	0.080264	3.267097	0.001	Adopt
Space support - > entrepreneur	0.145464	0.144271	0.083789	0.083789	1.736082	0.084	Adopt
Training support - > global	0.190104	0.188437	0.100004	0.100004	1.900959	0.059	Adopt
Training support - > entrepreneur	0.010234	0.009067	0.10902	0.10902	0.093877	0.925	Reject
Global - > non-financial	0.617304	0.624156	0.060973	0.060973	10.124179	0.000	Adopt
Global - > financial	0.3146	0.321724	0.083142	0.083142	3.783914	0.000	Adopt
Entrepreneur - > non-financial	0.181681	0.173092	0.070677	0.070677	2.570567	0.011	Adopt
Entrepreneur - > Financial	0.286835	0.281596	0.083094	0.083094	3.451912	0.001	Adopt
Commercialization - > Global	0.163389	0.162219	0.09654	0.09654	1.692451	0.092	Adopt
Commercialization - > Entrepreneur	−0.221412	−0.217122	0.12531	0.12531	1.766919	0.079	Reject
Acceler - > global	0.186022	0.188708	0.101723	0.101723	1.828716	0.069	Adopt
Acceler - > entrepreneur	0.336043	0.333038	0.120606	0.120606	2.786295	0.006	Adopt
Loan - > global	−0.108429	−0.108247	0.089104	0.089104	1.216881	0.225	Adopt
Loan - > entrepreneur	0.190663	0.190705	0.106004	0.106004	1.798629	0.073	Adopt
Consulting - > Global	0.227512	0.218929	0.112181	0.112181	2.028077	0.044	Adopt
Consulting - > Entrepreneur	0.317206	0.308242	0.111037	0.111037	2.85676	0.005	Adopt
Event support - > global	−0.224281	−0.215631	0.117643	0.117643	1.906457	0.058	Reject
Event support - > entrepreneur	−0.144376	−0.135447	0.11384	0.11384	1.268237	0.206	Reject

5 Conclusion

This study conducted an empirical study on the factors affecting corporate performance in this global start-up policy support. This study also identified the characteristics of start-up and this global start-up, and designed a research model based on previous studies on existing company support policies. The hypothesis is verified through empirical analysis, and the results are analyzed as follows:

First, the hypothesis on education support, commercialization support, loan support, and event support was rejected as for the variables affecting on entrepreneur orientation among policy support, and the hypothesis on space support, accelerating support, and consulting support were adopted. Entrepreneurial orientation can be interpreted as having little effect on funding, group education, and one-off events, and is influenced by space support and one-on-one professional support.

Second, all the hypotheses were adopted except the event support among the policy support factors which affects the global orientation. This means that in the case of start-ups that are relatively small and resource-insufficient, the government's policies on the global will be of great help in making this global start-up, and it is possible to prove that increasing the policy for the global is effective also for future start-up policy support.

Third, the hypothesis that the entrepreneurial orientation has a positive effect on the financial and nonfinancial performance of the firm was adopted. In particular, the non-financial performance of a company takes a large part in satisfaction with global performance, thus we can expect that the higher the entrepreneurial orientation, the higher the performance of global companies. Therefore, it can be interpreted that a firm having developed entrepreneurial orientation through support of space and one-on-one experts among entrepreneurship support policies has a positive impact on financial and non-financial performance.

Fourth, all hypotheses that global orientation affects firm's financial and non-financial performance are adopted. In particular, it has been examined that the higher the degree of global orientation, the higher the correlation with the non-financial performance of the firm. It is analyzed that this is because the higher the proportion of non-financial performance of the company, the higher the global satisfaction. Based on this, if the government's policy affecting the global orientation is supported to the start-up, it can be said that the start-up performance is also enhanced.

References

1. Oviatt, B.M., McDougall, P.P.: Toward a theory of international new ventures. J. Int. Bus. Stud (1994)
2. Oviatt, B.M., Shrader, R.C., McDougall, P.P.: The internationalization of new ventures: a risk management model. Adv. Int. Management. **16** (2004)
3. Han, U.-S., Lee, B.-K.: A study on the efficient management strategies for business incubator center in Korea. J. Bus. Educ. **6** (2003)

4. Cho, J.-H., Park, K.-Il.: A study on the effects of operating characteristics of business incubator on the firm's performance, Korean Acad. Soc. Acc. **13**(3) (2008)
5. Na, S.-G.: A study on the structural relationship among entrepreneurial characteristics, success factors and performances of small business start-up founders. Management Info. Syst. Rev. **35**(4) (2016)
6. Lee, Y., Hong, K., Jeong, Y., Park, S.: Developing entrepreneurship education strategy for venture's successful startup. Korea Soc. Learning Performance. **18**(1) (2016)
7. Kim, Y.J.: A study on building program development model of entrepreneurship education for adults. J. Vocational Educ. Training. **6**(1) (2003)
8. Mun, W.: Study on direction of development of business consulting for small and medium enterprises (2010)
9. Kim, C.-K.: A study on the operating performance factor of the BI's tenant. Korean Assoc. Bus. Adm. **2011**(1) (2011)
10. Sung H.J., Hae, R.K.: An exploratory study on success factors in small business startups. J. Small Bus. Innovation. **4**(2) (2001)
11. Han, J.H.: A Study on Activating Factors of Venture Entrepreneur in Korea (2008)
12. Lumpkin, G.T., Dess, G.G.: Clarifying the entrepreneurial orientation construct and linking it to performance. Academy Management Rev (1996)
13. Naman, J.L., Slevin, D.P.: Entrepreneurship and the concept of fit. Strateg. Management J. **14**, 137–153 (1993)
14. Coviello, N., Munro, H.J.: Growing the entrepreneurial firm. Euro. J. Marketing. **27**(7), 49–61 (1995)
15. Cavusgil, S.T., Zou, S.: Marketing strategy: performance relationship: an investigation of the empirical link in export market ventures. J. Marketing. **58**, 1–21 (1994)
16. Knight, G.A.: Cross-cultural reliability and validity of a scale to measure firm entrepreneurial orientation. J. Bus. Venturing. **12**, 213–225 (1997)
17. Eppink, D.J., Van Rhijn, B.M.: The internationalization of Dutch insurance companies. Long Range Plann. **21**(5), 54–60 (1988)
18. Ramaswamy, R., Rowthorn, R.: Deindustrialization: causes and implications (1997)
19. Knight, G., Koed Madsen, T., Servais, P.: An inquiry into born-global firms in Europe and the USA. Int. Marketing Rev. **21**(6), 645–665 (2004)
20. Kuivalainen, Q.,Sundqvist, S., Servais, P.: Firms' degree of born-globalness, international entrepreneurial orientation and export performance. J. World Bus. (2007)
21. Ari, J., Niina, N., Kaisu, P., Sami, S.: Strategic orientations of born globals: do they really matter? J. World Bus. **43**(2), 158–170 (2008)
22. Kim, G.: (Amos 18.0) Structure Equation Model Analysis (2010)
23. Lee, H.: Data Analysis Using SPSS (2013)
24. Kim, G.S.: A Study on the service quality strategy of university education. Korean Soc. Q. Management. **2010**(1) (2010)
25. Fornell, C., Larcker, D.F.: Evaluating structural evaluation models with unobservable variables and measurement error. J. Marketing Res. **18**(1), 39–50 (1981)

Design of the Model for Indoor Location Prediction Using IMU of Smartphone Based on Beacon

Jae-Gwang Lee, Seoung-Hyeon Lee and Jae-Kwang Lee

Abstract IPS (Indoor positioning system) is a system that measures the user's position in the room. Since IPS can't use GPS (Global Positioning System), various researches are under way focusing on indoor location accuracy. IPS may also be unable to measure indoors because of signal loss, blind spots, etc. To solve this problem, Beacon's RSSI signal is linearized using BITON algorithm and Kalman filter is applied. In addition, the position is predicted even when the signal is lost by measuring the instantaneous direction and the moving distance using the sensor of the smartphone. Therefore, in this paper, we propose a room location prediction model that can improve user's position accuracy and detect user's position in case of signal loss using Beacon and smartphone sensor.

Keywords Beacon · Kalman filter · Geomagnetic sensor · IPS
Indoor location prediction

1 Introduction

Due to the recent enlargement of indoor space, outdoor activities have become possible indoors, and the time spent living indoors has increased [1]. Also, as indoor structure becomes complicated, the importance of indoor location service is increasing. Outdoors, outdoor positioning technology based on GPS is being used. However, there is a disadvantage that GPS can not be used indoors. Therefore, indoor positioning technology using Zigbee, Wi-Fi, Bluetooth and UWB is being studied [2].

J.-G. Lee (✉) · J.-K. Lee
Department of Computer Engineering Hannam University, Daejeon, Korea
e-mail: jglee@netwk.hnu.kr

J.-K. Lee
e-mail: jklee@hnu.kr

S.-H. Lee
Information Security Research Division, ETRI, Daejeon, Korea
e-mail: duribun2@gmail.com

R. Lee (ed.), *Software Engineering Research, Management and Applications*, Studies in Computational Intelligence 789, https://doi.org/10.1007/978-3-319-98881-8_11

The WPS (Wi-Fi Positioning System) measures the location using the RSSI (Received Signal Strength Indication) of the wireless AP. However, in real space, there is a difference in accuracy depending on the number of wireless APs, and error increases in real space depending on signal interference, obstacle, and device characteristics [3].

UWB (Ultra Wide Band) utilizes a scheme that transmits signals over a very wide frequency band for low-power, large-capacity data transmission at short distances. Although relatively accurate position measurement is possible than WLAN (wireless local area network), the positioning time is delayed by the flight time of the reflected signal. In addition, since a wide frequency band is used, an error due to communication interference with other communication systems is large [4].

The most representative method of using Bluetooth is Beacon. Beacon is attracting attention as an indoor positioning technology with BLE technology. It is possible to position the user based on RSSI value like WLAN. However, due to the size of the signal, it has problems such as fading, signal interference, and loss of signal [5].

Thus, existing positioning technology has a disadvantage that accuracy is not high. Therefore, it is limited to areas where the IPS can't be applied to various fields and the position is not precise in coupon issuance, event notification, discount information, and the like.

For example, in the case of a car navigation system, the location of a car is continuously provided even in a tunnel or an underground driveway. Thus, it is necessary to provide continuous position even if the IPS.

There is PDR (Pedestrian Dead Reckoning) technology which is a method of measuring the position using the sensor of smartphone. The PDR technique estimates the position using information on the moving speed and the traveling direction. PDR uses acceleration sensor, gyroscope and geomagnetic sensor of smart phone. It is possible to measure the position of the user by using only the smartphone without building the infrastructure. However, due to the cumulative error, the error becomes larger as time passes and the accurate prediction is impossible [6].

In this paper, we propose a user location prediction model which can always show the user's position even in case of loss of signal by using Kalman filter with BITON based on Beaocn and sensor value of IMU of smart phone. This paper aims to study Beacon—based indoor location prediction technique for indoor location service as the interior becomes larger. Thus, this paper is composed as follows. In Sect. 2, techniques and algorithms for indoor location prediction are analyzed. In Sect. 3, the Kalman filter with BITON and the geomagnetism, gyrocope, and acceleration sensor of smartphone are used to design a location prediction model for indoor prediction. Finally, Sect. 4 describes the results and discusses future research.

2 Related Research

2.1 Beacon-Based Indoor Positioning Model

Beacon is a data communication technology that can exchange information between terminals based on BLE (Bluetooth low energy) which is a low power based Bluetooth 4.0 technology. Because it is based on BLE, it consumes very little power and is used in various fields. Recently, Beacon has been widely used for providing art information, merchandise advertisement, coupon issue, etc. in a museum. In order to use Beacon's RSSI value as an indoor location measurement technique, 'indoor location positioning system for providing indoor location based service', 'indoor positioning and moving distance measurement system using beacon and acceleration sensor.' 'Design and Application of Integrated Positioning System for Location Based Services' has been actively researched [7].

In addition, Beacon has features that are easy to install in existing buildings and low in construction cost. Therefore, this paper discusses the indoor location prediction technique based on Beacon.

2.2 Pattern-Based Indoor Location Prediction Model

A study on location prediction of moving objects constructs a space—time index on space—time data of moving objects. Based on these indexes, we have focused on predicting the location of future data. TPR-tree is a technique that can quickly search for the near future position considering speed and direction [8]. This technique excludes long-term signal loss. It also shows the difference from the movement of the actual moving object.

RMF (Recursive Motion Function) is proposed as a method to predict the position by predicting more recent movements [9]. Since this technique is represented by a nonlinear function, it shows a lot of errors when using Beacon based signals.

A method of predicting the future position of a moving object is based on the trajectory pattern mining technique [1]. This technique provides relatively accurate position estimates.

This pattern-based location prediction model requires a large amount of data and needs to be built in advance. Also, there is a disadvantage that environment setup time is consumed in proportion to high accuracy.

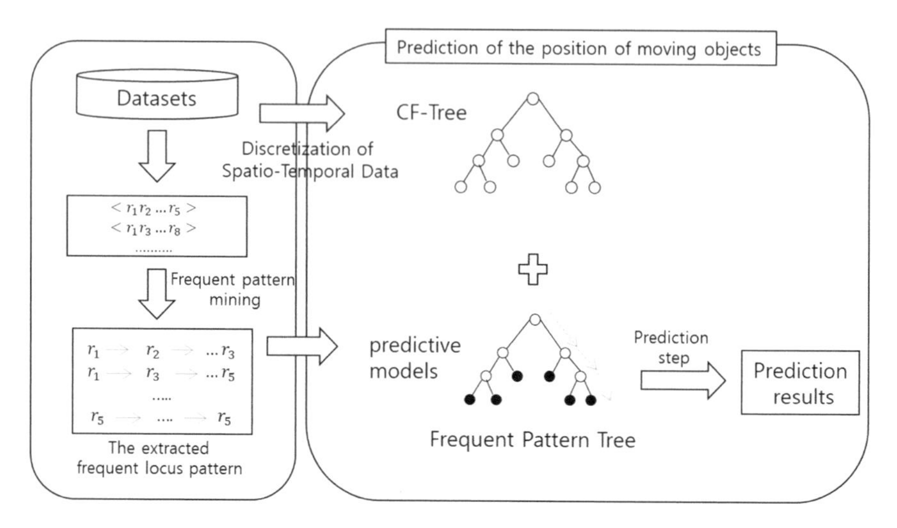

Fig. 1 Structure of moving object location prediction system based on frequent trajectory pattern

2.3 Indoor Location Prediction Model Based on Frequent Pattern

The moving object provides important information for predicting the location of the moving object. Using this information, a location prediction model with high accuracy can be constructed. The frequent trajectory pattern can solve problems such as inefficiency of Mining process and deterioration of Mining performance caused by inefficient space-time property. Figure 1 shows the overall structure of a moving object position prediction system based on frequent trajectory patterns.

First, the spatiotemporal data is idealized and the pattern is mined to derive a frequent trajectory pattern. Since the amount of derived data is large, it is possible to predict moving objects using CF-Tree and frequent pattern tree structure [10]. However, this model has poor prediction accuracy. It also has the disadvantage that the accuracy is lower when there is little data or the signal is lost.

2.4 PDR Based Positioning

PDR (Pedestrian Dead Reckoning) is a method of estimating the moving speed and traveling direction of an object to measure the position of the moving object [11]. In indoor location measurement using PDR, the sensor of the user's smartphone is used. The PDR method is a method of estimating the movement of a pedestrian by analyzing the number of steps, stride and direction, and analyzing the position and the like [12]. For PDR, the accuracy over short time is high. However, there is a

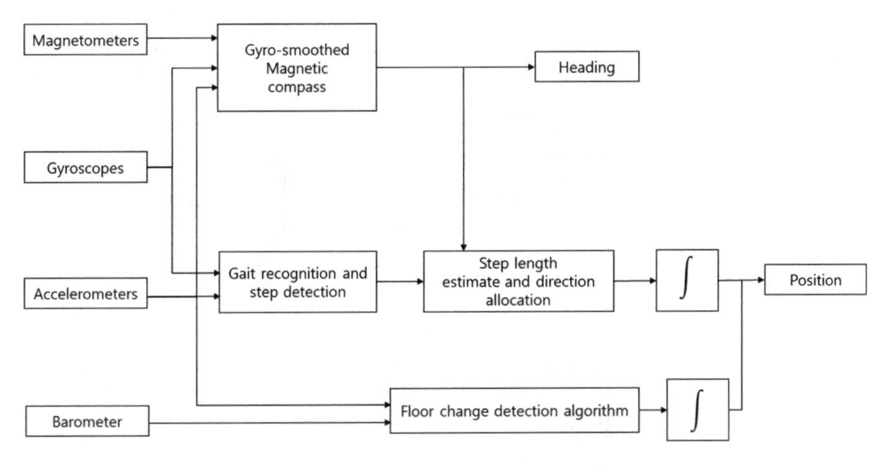

Fig. 2 Positioning process using PDR [13]

disadvantage that the error accumulates over time and the error becomes large after a long time [13].

To solve this problem, there is a method of measuring the user position using Beacon's RSSI signal and correcting the position using PDR [14]. However, this method has a problem of showing a large error even though it is possible to measure the signal loss. Figure 2 is a positioning process using PDR.

2.5 *BITON Algorithm*

The BITON algorithm is an algorithm for converting nonlinear data into linear data. Nonlinear data has irregular characteristics. The signal of GPS has characteristics of nonlinear data. For this reason, an error occurs when the position measurement is performed with the GPS signal. To solve this problem, we use the BITON algorithm to convert nonlinear data to linear data.

Since Beacon is nonlinear data, signal deformation occurs due to signal interference, fading, and so on. To solve this problem, it is necessary to apply the BITON algorithm to convert nonlinear data into linear data [14] (Fig. 3).

2.6 *Kalman Filter*

The Kalman filter is a recursive filter that is used to track the linear state of the noise. The Kalman filter is based on progressive measurements over time and is often used in position estimation systems because it provides more accurate results than only using the results at the moment of measurement. The following Fig. 4 shows the flow chart of the Kalman filter.

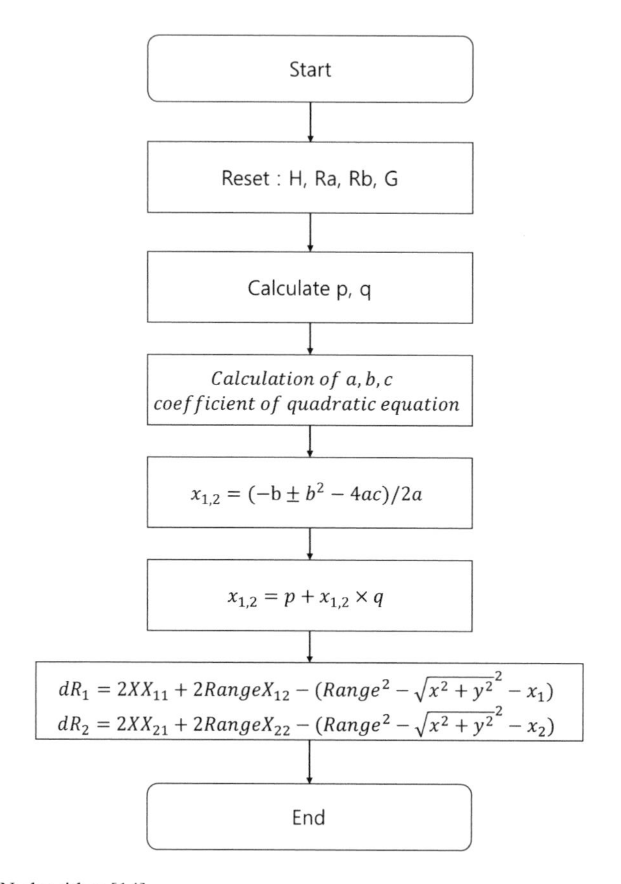

Fig. 3 BITON algorithm [14]

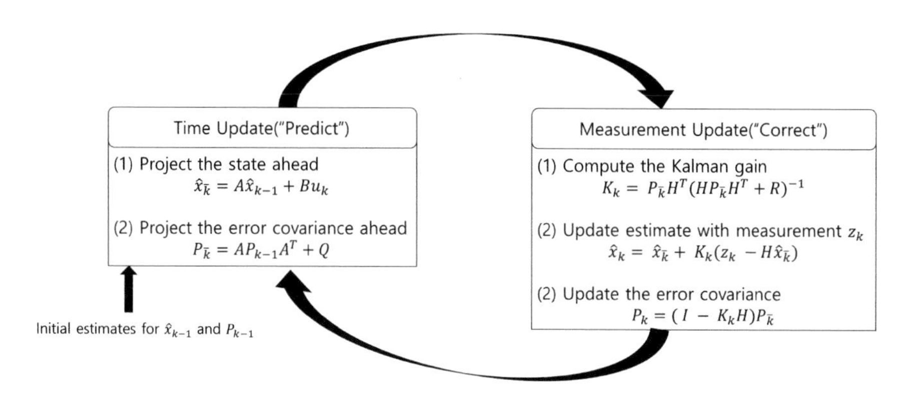

Fig. 4 Flow chart of the Kalman filter [15]

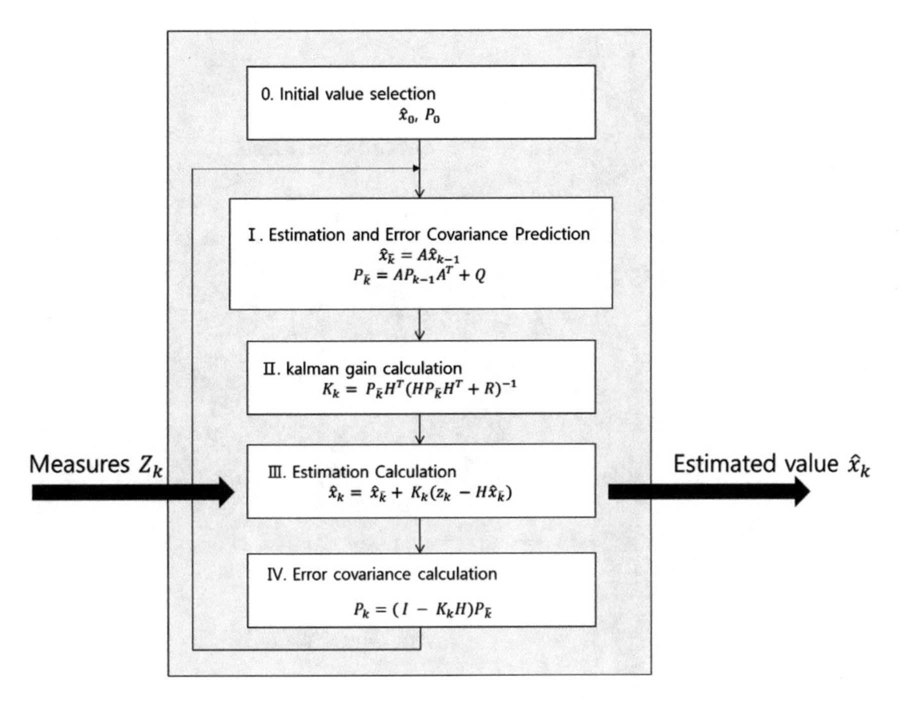

Fig. 5 The steps of the Kalman filter [15]

The Kalman filter can be largely divided into 'Predict' and 'correct'. If you subdivide it, you can divide it into 4 levels.

Step 1: This step predicts the estimated value and the error covariance as a prediction step.
Step 2: Calculate the gain of Kalman fiter.
Step 3: Calculate the estimated value using the input value.
Step 4: Calculate the error covariance.
This is illustrated as follows [15] (Fig. 5).

3 Design

In the indoor position measurement of the user, a signal loss, a shadow area, and the like occur. In such a situation, it is necessary to measure the position of the user. To solve these problems, the proposed model is expressed as follows (Fig. 6).

In this paper, we can divide into two situations. The position measurement is when the signal of the beacon is measured, and the position prediction means when the signal of the beacon is not measured.

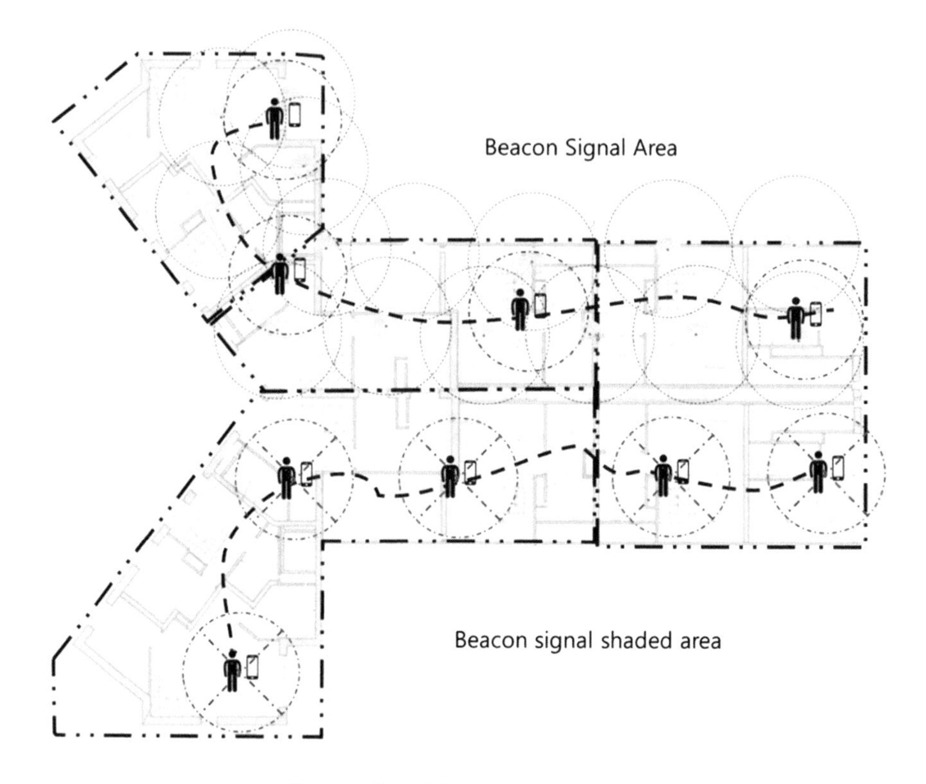

Fig. 6 Location prediction configuration

3.1 Position Measurement

Position measurement is a situation where Beacon's signal is measured normally. The following figure shows the flow chart for position measurement (Fig. 7).

The sequence of position measurement is as follows.
1. Beacon's signal is periodically broadcast.
2. Check whether the beacon signal is on the smartphone.
3. When the signal is confirmed, the beacon ID and RSSI are transmitted to the BITON algorithm.
4. Set the initial coordinates using Beacon's signal.

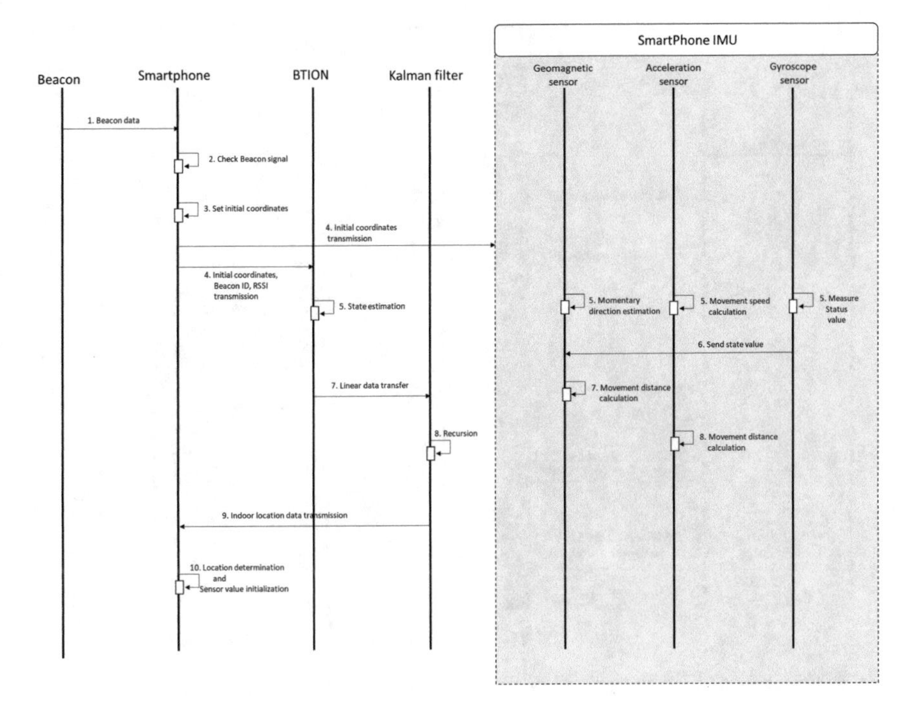

Fig. 7 Flow chart for position measurement

5. Transmit the initial coordinates, beacon ID, and RSSI to the BITON algorithm. The initial coordinates are transmitted to the IMU of the smartphone for position prediction.
6. Linearize the Beacon signal with the BITON algorithm.
SmartPhone's IMU measures sensor values for position prediction.
7. Pass the linearized coordinates to the kalman filter.
8. Estimate the position by applying the linearized coordinates to the Kalman filter.
9. Deliver the estimated location to the IMU of the smartphone and smartphone.
10. Confirm the position and initialize the value of IMU.

Position measurements can use Beacon's signal, but signal loss, shadow areas can occur at any time. Therefore, it is necessary to continuously estimate the sensor value for the position prediction even in the IMU of SmartPhone (Fig. 8).

3.2 *Location Prediction*

Location prediction is when there is no Beacon signal due to loss of signal or shadow area. In this case, the position of the user is predicted by using the sensor of the smartphone (geomagnetism sensor, acceleration sensor, gyroscope sensor). First, the

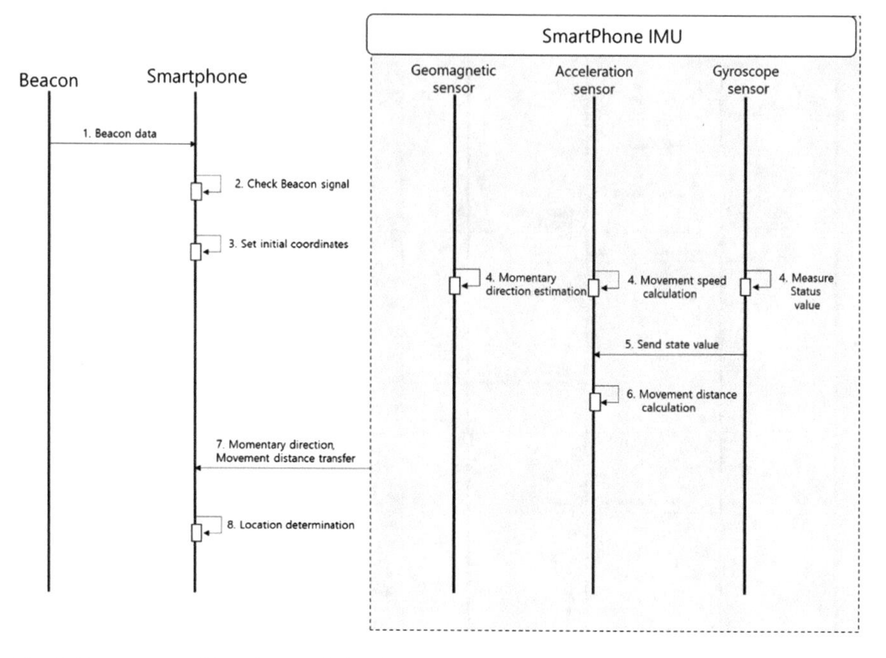

Fig. 8 Progression of location prediction in case of loss of beacon signal

direction of the user's instantaneous movement is estimated using the geomagnetic sensor. In addition, the movement speed is calculated using an acceleration sensor and a gyroscope sensor. And estimates the position by calculating the movement distance of the user using the estimated movement direction and the calculated movement speed. Figure 9 shows the progression of the location prediction when the beacon signal is lost.

The location prediction can be divided into five stages: an initial positioning step, a momentary direction estimating step, a moving speed estimating step, a moving distance calculating step, and a position fixing step.

The initial position setting step is to set the initial position of the user. If the beacon signal is measured and then turned off, the final position determined by using the beacon signal becomes the initial position. If you start without a beacon signal, you will use the initial position by specifying the initial position. In order to predict the position, it is necessary to initialize the last position to the initial position.

Secondly, when the direction of the user changes to the instantaneous direction estimation step, the geomagnetic sensor measures the direction using the magnetic field of the earth, so that the instantaneous direction can be measured. Inside the geomagnetic sensor, there are three sensors in the X, Y and Z axis directions that can measure the magnetic field strength. The magnetic field direction is measured by the sum of the outputs of these sensors. The following equation is a formula for finding the forward path.

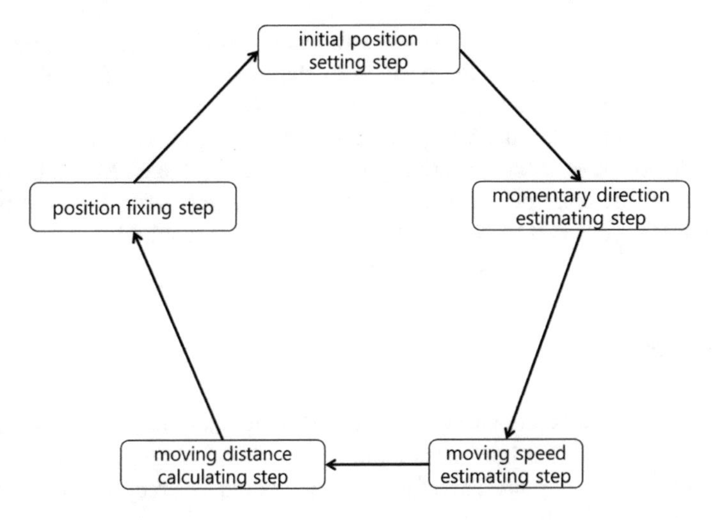

Fig. 9 Position prediction step

$$\psi = \tan^{(-1)}\left(\frac{Y_f g \cos\varphi - z_f g \sin\varphi}{X_f g \cos\theta - Y_f g \sin g\theta \sin\varphi - Z_{fg} \sin\theta \sin\varphi}\right) \tag{1}$$

ψ is the traveling direction, X_{fg}, Y_{fg}, Z_{fg} are the values of the geomagnetic sensor, θ is the roll, and φ is the pitch.

Third, the speed estimation step is divided into two. If the beacon signal is present and then disappeared, the movement speed can be estimated using the measured current position and the previous position. Equation (2) is as follows. If the log for the beacon signal is not present, the acceleration of the acceleration sensor is integrated as shown in Eq. (3) to estimate the movement speed.

$$\mathrm{v} = \frac{\sqrt{(x_2 - x_1)^2 + (y_2 - y_1)^2}}{t} \tag{2}$$

$$\mathrm{v} = \int(\vec{a})dt \tag{3}$$

v is velocity, s is distance, t is time, x_{ij}, y_{ij} are coordinates, and $\vec{a}$ is acceleration. The moving distance can be estimated by estimating the moving speed of the user.

The fourth step is to obtain the moving distance. The moving distance is obtained by Eq. (4).

$$\mathrm{s} = \mathrm{v} \cdot \mathrm{t} \tag{4}$$

Finally, the final step is to confirm the location. The position determination step is a step of coordinate the position of the user using the movement direction and the movement distance.

Position measurement and position estimation can't operate independently. If the beacon signal is present, the position of the user is indicated by the position measurement logic. However, in the case of location prediction logic, it operates in the IMU of the smartphone to prepare for loss of signal and shadow areas. In the shadow area, the indoor location prediction logic operates and initializes the IMU value of the smartphone when the signal is caught. After that, the position measurement logic needs to be activated and resetting is required.

4 Conclusion

As the size of indoor is increased, the time for living indoor has been increased and the structure of indoor has become complicated, so that the importance of the indoor location service is important. Therefore, indoor location accuracy, loss of signal, and location prediction in shaded area are necessary.

In this paper, we have linearized the nonlinear Beacon signal using BITON. The model is designed to improve the position accuracy by applying the linearized signal to the Kalman filter. We also proposed a prediction model using IMU of smartphone in signal loss or shadow area. The gyroscope sensor can be used to estimate the instantaneous direction and the acceleration sensor to estimate the user's movement speed. We propose a model that can estimate the user's position using the moving distance and the instantaneous direction by calculating the moving distance using the moving speed.

The predictive model proposed in this paper can represent the indoor position in any situation. Therefore, it is expected that indoor location service, which was applied to limited service such as advertisement, coupon, etc., can be applied to various fields. However, it is necessary to demonstrate excellence by implementing a predictive model.

Therefore, it is necessary to analyze the position accuracy and prediction accuracy of the position prediction model by implementing the prediction model proposed in this paper.

Acknowledgements This research was supported by Basic Science Research Program through the National Research Foundation of Korea (NRF) funded by the Ministry of Education (NRF-2017R1D1A3B03036130).

References

1. Salas, A.C.: Indoor Positioning System based on Blutetooth low Energy. Bachelor Thesis, Universitat Politecnia de Catalunia Barcelonatech, Barcelona (2014, June)
2. Zhu, J., Luo, H., Li, Z.: RSSI based bluetooth low energy indoor positioning. In: 2014 International Conference on Indoor Positioning and Indoor Navigation (2014, October)

3. Bahl, P., Padmanabhan, V.N.: RADAR: an in-building RF-based user location and tracking system. In: Proceedings of 19th Annual Joint Conference of the IEEE Computer and Communications Societies (INFOCOM'00), vol. 2, pp. 775–784 (2000, March)
4. Fang, L., Antsaklis, P.J., Montestruque, L.A., McMickell, M.B., Lemmon, M., Sun, Y., Fang, H., Koutroulis, I., Haenggi, M., Xie, M., Xie, X.: Design of a wireless assisted pedestrian dead reckoning system—the NavMote experience. IEEE Trans. Instrum. Measur. **54**(6), 2342–2358 (2005)
5. He, S., Chan, S. H. G., Yu, L., Liu, N.: Fusing noisy fingerprints with distance bounds for indoor localization. In: 2015 IEEE Conference on Computer Communications (INFOCOM), IEEE (2015)
6. Rai, A., Chintalapudi, K.K., Padmanabhan, V.N. Sen, R.: Zee: zero-effort crowd sourcing for indoor localization. In: Proceedings of the 18th Annual International Conference on Mobile Computing and Networking, ACM (2012)
7. Salas, A.C.: Indoor positioning system based on bluetooth low energy. A Degree's Thesis Submitted to the Faculty of the Escola Tècnica d'Enginyeria de Telecomunicació de Barcelona Universitat Politècnica de Catalunya (2014, June)
8. Tao, Y., Papadias D., Sun, J.: The TPR*-tree: an optimized spatiotemporal access method for predictive queries. In: Proceedings of the 29th International Conference on Very Large Data Bases, pp. 790–801 (2003)
9. Tao, Y., Faloutsos, C., Papadias D., Liu, B.: Prediction and indexing of moving objects with unknown motion patterns. In Proceedings of the 10th ACM SIGKDD International Conference Knowledge Discovery and Data Mining, pp. 611–622 (2004)
10. He, S., Gary Chan, S.-H.: Wi-Fi fingerprint-based indoor positioning: recent advances and comparisons. IEEE Commun. Surv. Tutorials (2015)
11. Rainer, M.: Indoor positioning technologies. Ph.D. Dissertation, Eidgenössische Technische Hochschule Zürich, Zürich, Switzerland (2012)
12. Li, X., Wang, J., Liu, C.: A Bluetooth/PDR integration algorithm for an indoor positioning system. Sensors **15**(10), 24862–24885 (2015)
13. Li, Z., Liu, C., Gao, J., Li, X.: An improved WiFi/PDR integrated system using an adaptive and robust filter for indoor localization. Int. Soc. Photogrammetry Remote Sens. Int. J. GeoInformation **5**(12), 224–239 (2016)
14. Cai, Y.D., Clutter, D., Pape, G., Han, J., Welge, M., Auvil, L.: MAIDS: Mining alarming incidents from data streams. In: SIGMOD 2004. Paris, France (2004, June)
15. Ali-Loytty, S., Perala, T., Honkavirta, V., Piche, R.: Fingerprint Kalman filter in indoor positioning applications. In: 3rd IEEE Multi-conference on Systems and Control (2009, July)
16. Fukuju, Y., Minami, M., Morikawa, H., Aoyama, T., DOLPHIN: an autonomous indoor positioning system in ubiquitous computing environment. In: Conference: Conference: Software Technologies for Future Embedded Systems (2003)

IoT Implementation of SGCA Stream Cipher Algorithm on 8-Bits AVR Microcontroller

Mouza Ahmed Bani Shemaili, Chan Yeob Yeun, Mohamed Jamal Zemerly, Khalid Mubarak, Hyun Ku Yeun, Yousef Al Hammadi and Yoon Seok Chang

Abstract Recently, the computing environment is changing due to the advent of resource-constrained devices, such as smart cards. That causes a limitation in providing some main feature to the low constraint computation devices such as security. The main purpose of this paper is to design a lightweight and secure stream ciphers for IoT to secure hardware and software that can fit constrain resources devices. Thus, we implement our proposed solution on 8 bits AVR microcontroller in order to study the required memory and speed. Also, two hardware algorithms which are Trivium and Grain are implemented and comparison provided between them and the proposed stream ciphers requirements. Also, our proposed SGCA algorithm proves to have less memory and time consuming than Trivium and Grain algorithms.

Keywords IoT · Stream cipher · AVR · Trivium · Grain

M. A. Bani Shemaili
CIS Division, HCT, Ras al Khaimah, UAE
e-mail: malahemaili@hct.ac.ae

C. Y. Yeun (✉) · M. J. Zemerly
ECE Department, Khalifa University of Science and Technology, Abu Dhabi, UAE
e-mail: cyeun@kustar.ac.ae

M. J. Zemerly
e-mail: jamal.zemerly@kustar.ac.ae

K. Mubarak
Dubai Men's College, HCT, Dubai, UAE
e-mail: kalhammadi1@hct.ac.ae

H. K. Yeun
NS Division, HCT, Abu Dhabi, UAE
e-mail: hyun.yeun@hct.ac.ae

Y. Al Hammadi
College of Information Technology, UAE University, Al Ain, UAE
e-mail: yousef-A@uaeu.ac.ae

Y. S. Chang
School of Air Transport and Logistics, Korea Aerospace University, Goyang, South Korea
e-mail: yoonchang@kau.ac.kr

R. Lee (ed.), *Software Engineering Research, Management and Applications*, Studies in Computational Intelligence 789, https://doi.org/10.1007/978-3-319-98881-8_12

1 Introduction

Nowadays, mobiles, smart cards, Internet of Things (IoT) are considered as integral part of our daily lives. These devices store and transmit our private confidential information, through unsecure communication channels. Thus, providing security of these communications is mandatory and out of the question. However, these devices face the challenge of resource constraint, thus designing a cipher with lightweight properties is always challenging that the designer needs to cope with the trade-off between achieving robust security with low memory requirement and enhanced performance [1]. Achieving a cipher design that can be light to fit these devices becomes a challenge by itself. A lightweight cryptography algorithm requires low hardware resources, high operation frequencies and very high security measures. This challenge can be achieved using stream ciphers. Stream ciphers are one of the approaches that are used in order to generate a symmetric cipher. Stream cipher algorithms encrypt one bit of the plain text at a time. They are famous of being fast, easy and need very low memory size in both of the software and hardware application [2]. These properties of stream cipher lead to be the best choice for the low computation power devices.

Surly all organization and forums agree on the need of strong security to secure their systems however, it is difficult to implement security algorithms within low computation devices. Besides, the securities of the most available systems are based on the outdated encryption algorithms that may works fine with high computation devices but have proven to be unsuitable for low computation hardware implementation.

Thus, this paper investigates the new proposed algorithms for low computation devices. The available algorithms are chosen from eStream project hardware profile which is Grain, Trivium and MICKEY. Also, the paper present our proposed security algorithm called SGCA. The SGCA algorithm designed to be a lightweight stream cipher algorithm for low constrains devices. The details of our algorithm can be found in [3].

The main aim of this paper is to investigate how efficient and lightweight our SGCA proposed stream cipher algorithm to be implemented on the small micro-controller with restricted memory resources. Thus, the paper illustrates the implementation on 8 AVR microcontroller of the SGCA algorithm, Grain, Trivium for the purpose of comparison. In this paper we did not implement MICKEY algorithm as the algorithm is more complicated than the rest.

The paper is organized as follow; Sect. 2 illustrates the related work. Section 3 explains our GSCA algorithm for completion. Section 4 shows the software implementation of the SGCA algorithm on 8 bits AVR microcontroller. Section 5 compares between the SGCA algorithm and Trivium and Grain algorithms. Finally, Sect. 6 concludes the paper.

2 Related Work

eStream is a well-known project start in 2004 that aims to encourage the designer to design stream cipher algorithms [4]. The eStream project consists of two categories which are software and hardware applications for restricted resources devices. Under the categories of the hardware there are Trivium and Grain stream ciphers. These two ciphers aim to be implemented on low constraint devices that are why they are designed with lightweight properties. Next subsection explains these two ciphers in more details.

2.1 Trivium

Trivium [5] is a hardware synchronous stream cipher that uses a key of 80 bits and IV of 80 bits. Trivium cipher internal state of 288 bits shift register denoted by (s1, …, s288) and updated at every clock cycle. It has reached the final portfolio of the eSTREAM project. The aim of the cipher is to provide a flexible speed and space.

(1) *The Initialization Process*: The initialization process start by loading 80 bits of the key and 80 bits of the IV into the shift register of the 288 bits initial state. The loading value of the key and IV will be zero to all except for s286, s287, and s288 as follow:

$$(\mathrm{K1}, \ldots, \mathrm{K80}, 0, \ldots, 0) \rightarrow (\mathrm{s1}, \mathrm{s2}, \ldots, \mathrm{s93})$$
$$(\mathrm{IV1}, \ldots, \mathrm{IV80}, 0, \ldots, 0) \rightarrow (\mathrm{s94}, \mathrm{s95}, \ldots, \mathrm{s177})$$
$$(0, \ldots, 0, 1, 1, 1) \rightarrow (\mathrm{s178}, \mathrm{s279}, \ldots, \mathrm{s288})$$

Then the initial state is rotated for 4 full clock cycles, however without generating any key stream bits.

(2) *Key Stream Generation*: The idea of Trivium is to replace the building block of the block cipher which is S-boxes and linear filters by stream cipher. Trivium replaces this part by data words that are arranged in clockwise direction and extracts 15 bits from specific bits that are used to update 3 bits which are (t1, t2, t3) of the state and produce 1 output bit which is denoted by zi of the stream cipher.
The cipher rotates the state bit and repeats the process again till it reaches the desirable key length that can reach 264. The inputs of the 15 bits are as follow:

$$\mathrm{t1} = \mathrm{s66} + \mathrm{s93} + \mathrm{s91} \oplus \mathrm{s92} + \mathrm{s171}$$
$$\mathrm{t2} = \mathrm{s162} + \mathrm{s177} + \mathrm{s175} \oplus \mathrm{s176} + \mathrm{s264}$$
$$\mathrm{t3} = \mathrm{s243} + \mathrm{s288} + \mathrm{s286} \oplus \mathrm{s287} + \mathrm{s69}$$

The output stream is the summation of the 3 t values: $\mathrm{zi} = \mathrm{t1} + \mathrm{t2} + \mathrm{t3}$.
Note that the '$\oplus$' and '+' operations stand for XOR and AND over GF (2) respectively.

(3) *Attacks on Trivium Cipher*: Trivium cipher can be attacked by the algebraic attack, and correlation attack as described in [6].

2.2 Grain

Grain [7] is designed for low constraint environments. The cipher is intended to be used in environments where gate count, power consumption and memory usage need to be very small. Grain design is based on two shift registers and a nonlinear filter function with a key size of 128 bits and an Initialized Vector IV of 128 bits.

The cipher consists of three main building blocks, namely an LFSR, an NFSR and a filter function. The content of the LFSR is denoted by si, $si + 1$, …, $si + 127$ and the content of the NFSR is denoted by bi, $bi + 1$, …, $bi + 127$. The feedback polynomial of the LFSR, $f(x)$ is a primitive polynomial of degree 128. Both of the LFSR and NFSR are updated with update function.

(1) *The Initialization Process*: The initialization process of the cipher starts by loading the NFSR with the key bits thus $bi = ki$ for $0 \leq i \leq 127$. Then the LFSR is loaded with IV bits thus $si = IVi$, for $0 \leq i \leq 95$. Then, in order to make sure that the LFSR will not initialize with zero state the remaining of the LFSR are filled with ones thus $si = 1$ for $96 \leq i \leq 127$. Then, the cipher is clocked 256 times without producing any output stream however; the output is fed back and XORed with the input of the LFSR and NFSR.

(2) *Key Stream Generation*: The feedback polynomial of the LFSR, $f(x)$ is a primitive polynomial of degree 128. It is defined as
$f(x) = x128 + x121 + x90 + x58 + x47 + x32 + 1$.
For the update function of the LFSR the following is applied:
$si + 128 = si + si + 7 + si + 38 + si + 70 + si + 81 + si + 96$.
The feedback polynomial of the NFSR, $g(x)$, is defined as
$g(x) = x117x115 + x111x110 + x88x80 + x101x96 + x67x63 + x61x125 + x44x60 + x128 + x102 + x72 + x37 + x32 + 1$.
For the update function of the NFSR the following is applied:
$bi + 128 = si + bi + bi + 26 + bi + 56 + bi + 91 + bi + 96 + bi + 3bi + 67 + bi + 11bi + 13 + bi + 17bi + 18 + bi + 27bi + 59 + bi + 40bi + 48 + bi + 61bi + 65 + bi + 68bi + 84$.
For the output stream zi, 9 variables bits are taken as input values for the Boolean function $h(x)$. the nine taps values are: $bi + 12$, $si + 8$, $si + 13$, $si + 20$, $bi + 95$, $si + 42$, $si + 60$, $si + 79$ and $si + 95$ corresponding to $x0$, $x1$, $x2$, $x3$, $x4$, $x5$, $x6$, $x7$ and $x48$ that are used in the following $h(x)$ filter function respectively.
$h(x) = x0x1 + x2x3 + x4x5 + x6x7 + x0x4x8$.
The output of the $h(x)$ filter function is masked with the bit bi from the NFSR in order to produce one bit of the key stream for each clock thus zi is produced as follow:
$zi = \sum j \in A\, bi + j\, h(x) + si + 93$, where $A = (2; 15; 36; 45; 64; 73; 89)$.

(3) *Attacks on Cipher*: The cipher is under attacks that aim to reconstruct the key stream sequence [8].

3 NEW SGCA Stream Cipher

SGCA based stream cipher combines both of the Linear Feedback Shift Register (LFSR) [9] and Feed Carry Shift Register (FCSR) [10] with Shrinking Genertaor (SG) [11] algorithm and used cellular automata (CA) [12] as an updated function. The combination of these elements is to ensure that the stream cipher can provide a secure cryptographic key that is hard to crack. The SGCA stream cipher operates in three processes, namely an initialization process, a key stream generation process and an update process.

3.1 Initialization Process

The first step is the initiation step that uses the key and the Initial Vector (IV) to produce the output stream. Both of the key and the IV have 128 bits in length. This step produces bits and updates the registers of the LFSR and FCSR each of which is 128 bits. The step will not produce any output and all the bits are fed back to the registers. Thus, after reaching 256 bits in the State this step is finished.

3.2 Key Stream Generation Process

The steps to generate the stream are as follow:

- Both of the FCSR and the LFSR are clocked together.
- If the output of the FCSR is 1 then the output of the state is the output of the LFSR at that state.
- If the output of the FCSR is 0 then the output stream is discarded. Figure 1 shows the key generation phase where Table 1 explains the rule.

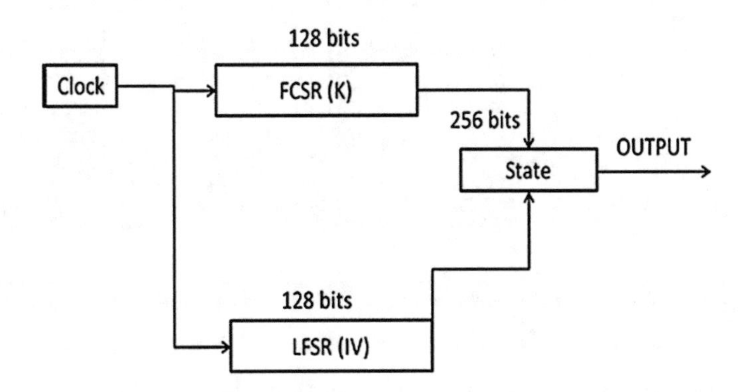

Fig. 1 The key generation main components

Table 1 Key stream generation rule

A(FCSR)	S(LFSR)	Z
1	0	Discards
0	1	Saves 0 in buffer
1	0	Discards
0	1	Saves 1 in buffer

3.3 Update Process

The update stage uses the Cellular Automata (CA) theory as a basis which is considered as a non-linear filter. In order to choose the rule all of the elementary CA rules are applied and changing the rules is applied till reaching the rule that can provide a balanced distribution between the zeros and ones with our stream cipher. The rule can randomly update the key and produce a new key each time. Each cell checks the surrounded cells and applies the following rules:

- If the cell is surrounded by 3 neighboring cells and if the number sum of cells with one = 2 or if the number sum of cells with zero = 3 then cell = 1.
- If the cell is surrounded by 4 neighboring cells and if the sum of cells with one = 2 or 3 or if the sum of cells with zero = 4 then cell = 1.
- Else cell = 0. Figure 2 explains the CA rules that applied on the key and IV in order to update them. For more information about the cipher stream please refer to [3].

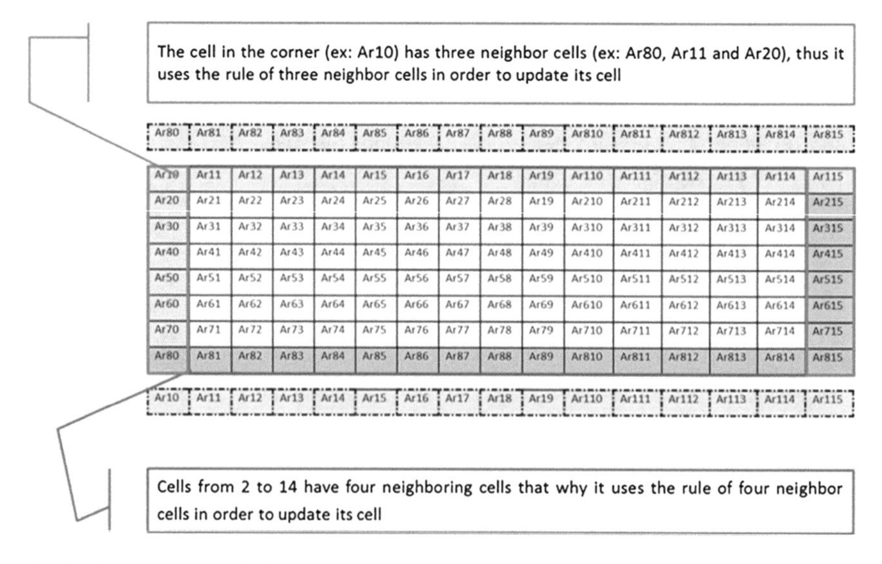

Fig. 2 CA rules that applied to updates the key and the IV

The first step is the initiation step that uses the key and the Initial Vector (IV) to produce the output stream. Both of the key and the IV have 128 bits in length. This step produces bits and updates the registers of the LFSR and FCSR each of which is 128 bits. The step will not produce any output and all the bits are fed back to the registers. Thus, after reaching 256 bits in the State this step is finished.

4 Statistical Test (St)

The Statistical Test (ST) is a way for the cryptographic cipher algorithm designer to decide if their cipher is producing a random like output sequence or not.

There are three approaches for the ST to decide if the produced output sequence is random or not. The first approach is the threshold values that calculate a statistic test, which is a statistic that determines the hypothesis basic of the test, for the output sequence and compares it to a threshold value.

However, this approach has a limitation since it considers the sequence that exceeds the threshold value as non-random even so it is random. The second approach is the fixed range approach which also calculates a test statistic for output sequence and then it compares it with a range if the value falls outside the range then it is considered as non-random. Also, this approach is considered not that flexible since in order to have a fixed range for the comparison then a significance level and acceptable range are pre-computed and if these levels are changed in the future then there is a need to compute the ranges values again. The most flexible ST is the third approach that uses a probability value. This approach calculates the test statistic for the output sequence and its equivalent probability value (P-value). The P-value is the probability of gaining a test statistic as larger than the one observed if the sequence is random. Usually the P-value is compared to a significance fixed value α thus if the output sequence P-value is larger than α it is considered random. The advantage of this approach is that it is not essential to have a measurement of the significance level just it needs to compute the P-value of the output sequence and see if it is $>\alpha$ to determine if the sequence is random or not.

One of the most known test suite for the ST is the NIST Statistical Test Suite [13] that uses the P-value in order to determine if the output sequence is random or not. The NIST Statistical Test Suite is based on hypothesis test to determine if the sequence output has the random characteristic or not. The ST is carried out by first computing the output sequence test statistic and computes P-value [0, 1] then compares if the P-value $>\alpha$ to determine if the sequence is random or not, where $\alpha; \in (0.001, 0.01]$. Within the NIST Statistical Test Suite there are 15 tests that are applied on the output sequence to determine the sequence randomness. The 15 tests are as follow:

1. Frequency Test (F): This test examines the frequency between ones and zeros to determine if there is an approximately equal number between generated

sequences. The best stream has equal probabilities between both of the zeros and ones.

2. Frequency Test within a Block (FTWB): In this test the stream sequence is divided into M blocks and the number of ones is examined between each blocks. The best sequences are those with the number of ones within M block is half.
3. Cumulative Sums Test (CS): This test examines the strip of zero in random walk by calculating the cumulative sum of adjusted $(-1, +1)$ digits in the sequence. By converting the stream ciphers input stream to $+-1$ by giving one value $=+1$ and zero value $=-1$. Then, discover if the cumulative sum is too large or small by adding the values of +1 and -1 depending on stream sequence. For example if the stream is 10110 then, it will be converted to $+1-1+1+1-1$ then the sequence will be:

   $S0 = 1$
   $S1 = 1 + (-1) = 0$
   $S2 = 1 + (-1) + 1 = 1$
   $S3 = 1 + (-1) + 1 + 1 = 2$
   $S4 = 1 + (-1) + 1 + 1 + (-1) = 1$

   Then by that the stream is compared to the behavior of the cumulative sum for random sequences if it contains too many ones or zeros.
4. Runs Test: This test finds out the number of changes in the stream sequences from ones and zeros. Then the test determines if the number of ones and zeros are expected for random sequence. The best stream sequence is that which has more changes within the sequence between zeros and ones.
5. Test for the Longest Run of Ones in a Block (TFLROB): This test finds out if the run of the longest run of ones within block of stream cipher is consistent with the expected longest run of ones with the random numbers.
6. Rank: This test finds out the rank of disjoint sub-matrices of the entire sequence. The purpose of this test is to check for linear dependence among fixed length substrings of the original sequence.
7. Discrete Fourier Transform (DFT): This test is used to find out the peak heights in the Discrete Fourier Transform of the sequence in order to detect periodic features such as repetitive patterns within the tested sequence that would indicate a deviation from the assumption of randomness.
8. Non-overlapping Template Matching Test (NOTM): The test observes if the stream is able to produce non periodic pattern or not by searching through the sequence of the availability of similar patterns. The test specifies a window for the specific pattern. If the pattern is found within the window then window is reset to the bit after the pattern that is found and if the pattern is not found then the window will slide by one bit position.
9. Overlapping Template Matching Test (OTM): The test detects the similar patterns that are produced from the stream cipher by searching through the sequence of the availability of similar patterns using a window with specific size. The test works in the same way as the previous test however in the case of finding the pattern then the widow slides one bit before resuming the search.

10. Universal Statistical Test (US): this test is used to find out if the sequence can be compressed without losing any information by finding out the number of bits between matching patterns which is considered as a measure that is related to the length of a compressed sequence.
11. Random Excursion Test (RE): This test is used to determine the number of cycles having exactly K occurrences within the cumulative sum random walk. The purpose of this test is to determine if the number of visits to a particular state within a cycle deviates from what one would expect for a random sequence.
12. Random Excursion Test Variant (REV): This test is about finding out the total number of times that occurs on a particular state in a cumulative sum random walk. The purpose of this test is to detect deviations from the expected number of visits to various states in the random walk.
13. Approximate Entropy (AE): This test finds out the frequency of all possible overlapping m-bit patterns across the entire sequence. The purpose of the test is to compare the frequency of overlapping blocks of two consecutive/adjacent lengths (m and m + 1) against the expected result for a random sequence.
14. Serial (S): This test finds out the frequency of all possible overlapping m-bit patterns across the entire sequence. The purpose of this test is to check if every m-bit pattern appears as frequently as every other m-bit pattern on average.
15. Linear Complexity Test (LC): This test examines the length of the register that can determine if the stream sequences are complex enough. Random sequences are considered by longer register since the size of the smallest register that generates a periodic sequence determines the linear complexity of it.

4.1 The NIST Statistical Test Results on the SGCA

The NIST statistical test is applied on the SGCA stream cipher output sequence to prove the randomness of the PPRNG output. As stated in the above section on ST, the stream can pass the test if the result value is more than 0.001 as shown in Fig. 3. The PPRNG is considered more random if the value is near 1 and non-random if the value is near or equal to zero.

The results of the ST on the SGCA algorithm can be shown in Fig. 4. Only 12 tests out of 15 are applied on the SGCA algorithm. The 3 remaining tests show negative values with the test table that why we determine to use the 12 tests results only. In

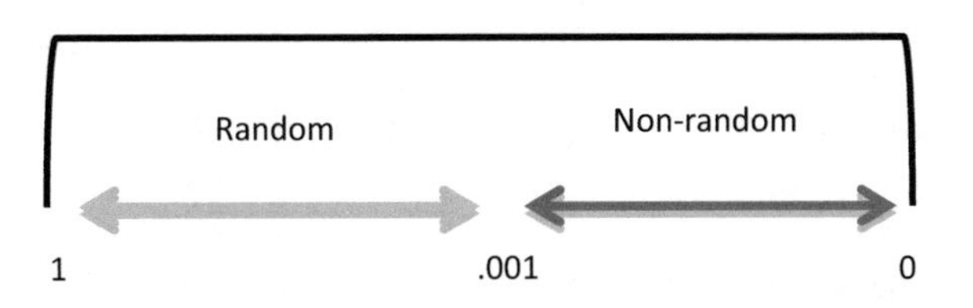

Fig. 3 Pass succeed rule

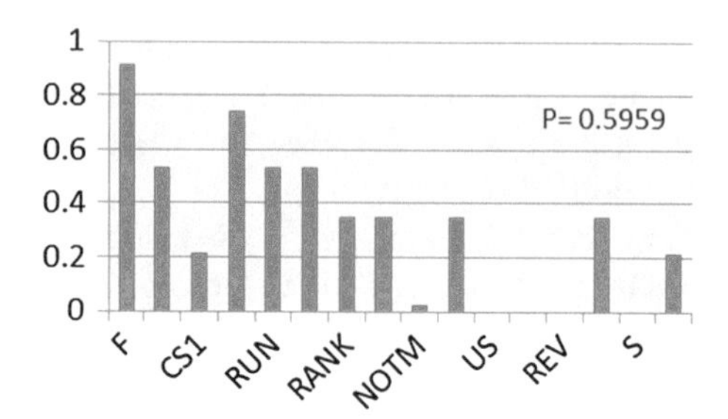

Fig. 4 ST results for the SGCA algorithm

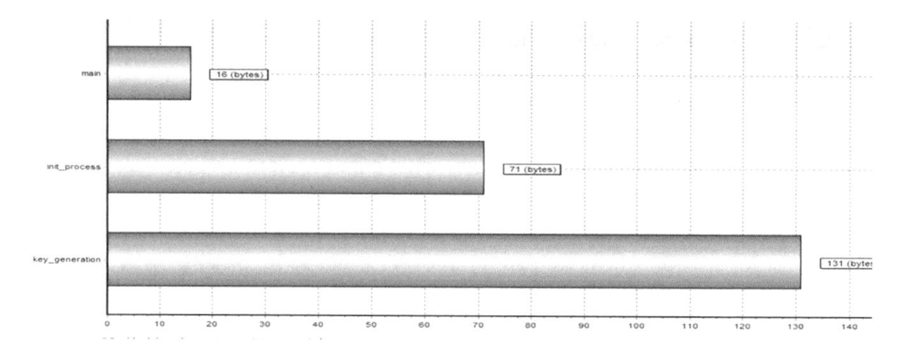

Fig. 5 Memory usage of the SGCA algorithm

over all the P value of the SGCA algorithm is 0.5959 which means that the cipher output is random. Next section explains the software implementation of the SGCA algorithm on 8 bit AVR microcontroller.

4.2 IoT Implementation of SGCA Stream Cipher

An 8 bit microcontroller can be used in low computation devices such as smart cards. Thus, we decide to use it as proof of lightweight of the proposed stream cipher for IoT. For our SGCA algorithm and other two algorithms the software implementations are realized using the C programming language since it is used to develop many applications nowadays. Besides, in term of performance and code size there is not much difference between C and other assembler languages.

For the development tool we used mikroC PRO for AVR [14] to program the SGCA, Trivium and Grain algorithms as well. The tool can facilitate our work since it is built in with many of hardware and software libraries. Besides the tool can provide statistical information about the code that can help us in understating the memory consumption of each algorithms. Figure 5 shows the memory consumption of the SGCA algorithm.

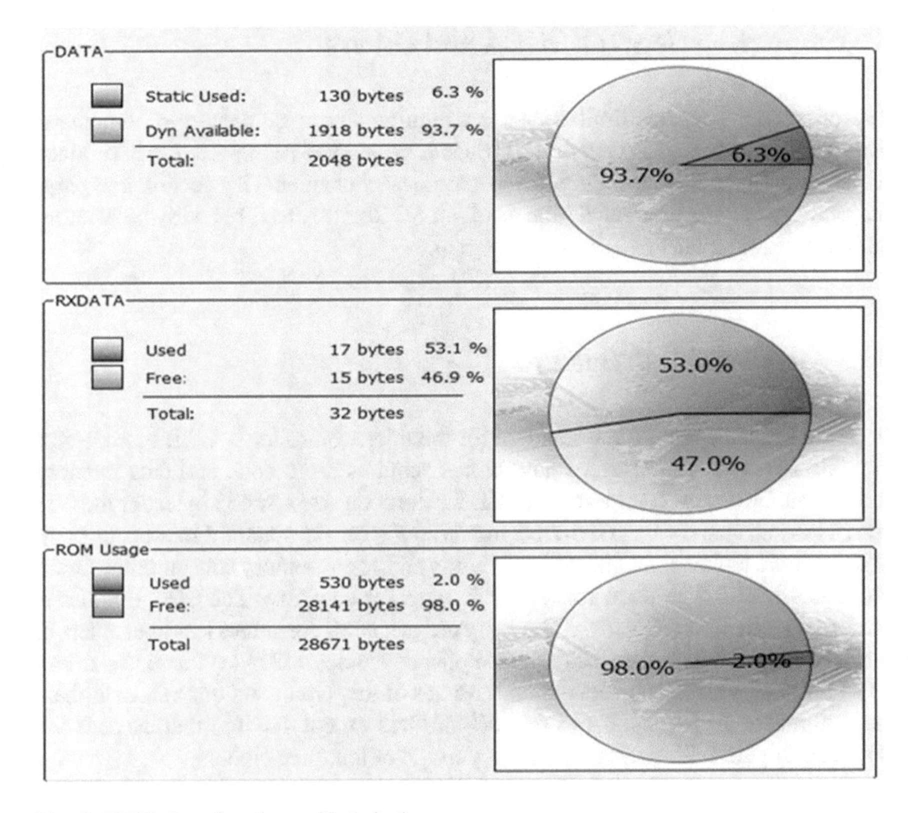

Fig. 6 SGCA three functions with their size

The memory is divided between dynamic and static variables. The static variables have fixed size where the dynamic variables do not have fixed size so the size increase or decrease depends on the variable needs. In case of memory usage the dynamic variables are much better and less costly since they end with the run time of the function unlike the static ones that ends at the end of running program and the memory can increased or decreased based on the code memory requirements [15]. The SGCA algorithm used 6.3% static and 93.7% dynamic variables.

On the other hand, Fig. 6 shows the size of each function on the SGCA cipher C code. The C code consists of 3 functions which are the main function, the initialization function and the key generation function. The SGCA main function consumes 16 bytes from the memory size where, initialization function consumes 71 bytes and the key generation function consumes 131 bytes. Next section compares between our SGCA cipher and Trivium and Grain ciphers based on memory and time consumption.

5 Comparing Between SGCA and Others

The microcontroller has limited memory requirement as compared to computers. Thus the memory requirement of the C code is considered as important criteria since it can increase or decrease the price of the microcontroller. This section compares the memory and the time requirements of our SGCA stream cipher with the memory and time requirements of Trivium and Grain.

5.1 Memory Requirements

In term of memory usage the standard for the microcontroller is 4 KB of code size and 256 bytes of data size. Trivium cipher requires more code and data memory usage than the others. It requires 1.67 KB for the code size and 433 bytes for the data size. The data size for the Trivium cipher is so big for the standard microcontroller. Where, Grain cipher comes next after Trivium cipher on memory consumption. Grain consumes 1.17 KB in code size and 263 bytes for data size. The SGCA cipher is efficient in terms of code and data memory usage since it consumes less than others in terms of code and data size. The SGCA cipher consumes 942 bytes for code size and 218 for the data size. The data memory usages of the, Grain and our SGCA ciphers are reasonable sizes for most of microcontrollers except Trivium that exceeds the data size. Figure 7 illustrates the memory usage of the three ciphers.

5.2 Time Requirements

Time is considered as one of the important criteria in order to evaluate the performance of the ciphers. The time of the initialization and key generation of each cipher are estimated. Estimated the time for the ciphers in both processes (initialization

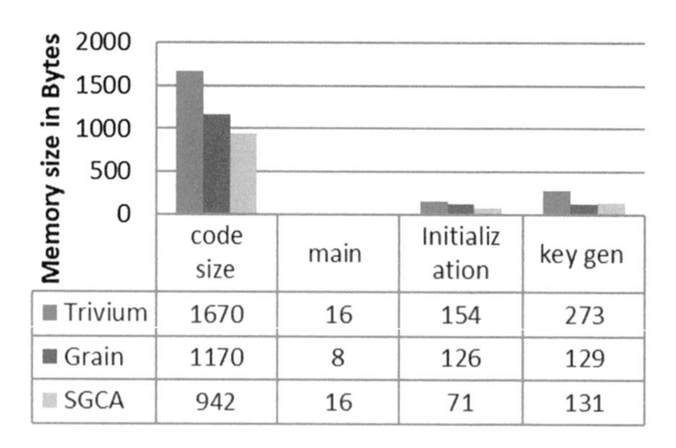

	code size	main	Initialization	key gen
Trivium	1670	16	154	273
Grain	1170	8	126	129
SGCA	942	16	71	131

Fig. 7 Memory usage for the three ciphers

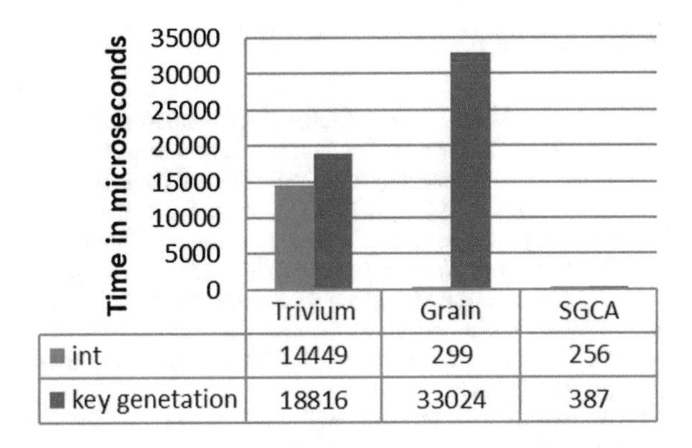

Fig. 8 Time requirements for the three ciphers

and key generation) are bounded with two parameters which are the algorithm structure such as the numbers of loops in the code and the selected target device. In this study, we used Atmega32 microcontroller with 10 MHz crystal frequency as a target device. For the execution time the SGCA cipher has the least time than others in the initialization and key generation time.

It consumes 256 microseconds for the initialization process and 387 microseconds for the key generation time which is much less than the key generation of the grain and Trivium to the percentage of 85% between Grain and SGCA and 47% between Trivium and SGCA. The executions time of the initialization process and key generation process for three ciphers is shown in Fig. 8.

6 Conclusion

In this paper we have presented the IoT implementation of our SGCA cipher. Also, we have implemented the Trivium and the Grain ciphers that belong to eStream project and reached the final stage in the hardware design. As we intend to convert our SGCA cipher into hardware design also a comparison between our SGCA ciphers and other hardware cipher become mandatory. From the C implementation we realized that the SGCA cipher fits the 8 bits AVR microcontroller and it is proof that better than other comparable ciphers such as Trivium and Grain. SGCA cipher can provide security to the low computation devices with low price as it need small memory space as we conclude from this paper. Also, the time consumption of SGCA is better than the Trivium and Grain ciphers as it consumes less time in both of the initialization and key generation processes. Thus, we recommend the use of SGCA cipher.

References

1. Robshaw, M.J.B.: Stream Cipher. RSA Tech. Report, TR-701, 2, 1–42 (1995)
2. Eisenbarth, T., Paar, C., Poschmann, A., Kumar, S., Usadel, L.: A survey of lightweight—cryptography implementations. In: IEEE CS and the IEEE CASS, pp. 522–533 (2007)
3. Bani Shemaili, M., Yeun, C. Y., Mubarak, K., Zemerly, M.J.: A novel hybrid cellular automata based cipher system for internet of things. In: Future Information Technology (FutureTech 2013), Lecture Notes in Electrical Engineering 276, pp. 269–276 (2013)
4. ECRYPT. eSTREAM: ECRYPT Stream Cipher Project, IST-2002-507932. Available at http://www.ecrypt.eu.org/stream/
5. De Canni'ere, C., & Preneel, B.: Trivium. In: New Stream Cipher Designs, 4986 of LNCS, pp. 84–97. Springer (2008)
6. Aumasson, J.-P., Dinur, I., Meier, W., Shamir, A.: Cube testers and key recovery attacks on reduced-round MD6 and trivium. In: FSE, LNCS. Springer (2009)
7. Martin, H., et al.: A stream cipher proposal: Grain-128. In: 2006 IEEE International Symposium on Information Theory. IEEE (2006)
8. Itai, D., et al.: An experimentally verified attack on full Grain-128 using dedicated reconfigurable hardware. In: Advances in Cryptology–ASIACRYPT 2011, pp. 327–343. Springer, Berlin (2011)
9. Alfke, P.: Efficient Shift Registers, LFSR Counters, and Long Pseudo-Random Sequence Generators. Xilinx, Tech. Rep., (Version 1.1) (1996)
10. Klapper, A., Goresky, M.: 2-adic shift registers. In: Fast Software Encryption (FSE 1993), Lecture Notes in Computer Science 809, pp. 174–178. Springer (1993)
11. Coppersmith, D., Krawczyk, H., Mansour, Y.: The shrinking generator. In: Proceedings of the Advances in Cryptology (CRYPTO 1993), LNCS, pp. 22–39 (1994)
12. Wolfram, S.: Cryptography with cellular automata. In: Advances in Cryptology: Crypto'85 Proceedings, LNCS 218, pp. 429–432. Springer (1985)
13. Bassham, L., et al.: A Statistical Test Suite for Random and Pseudorandom Number Generators for Cryptographic Applications, pp. 800–822. NIST SP (2010)
14. MIKROELEKTRONIKA: mikroC PRO for AVR, Dec., (2008). http://www.mikroe.com/downloads/get/300/mikroc_pro_avr_manual_v100.pdf
15. PatilÀ, N.V., Irabashetti, P.S.: Dynamic memory allocation: role in memory management. Int. J. Curr. Eng. Technol. **4**(2), 531–535 (2014)

A Study on Upgrading Non-urban Areas-Using Big Data the Case of Hwang Ze and Danggok Districts

Yong Pil Geum

Abstract This study examines areas with potential future growth in resident populations through political implication from population expansion in non-urban areas. Using big data Hwang Ze and Danggok districts in Gyeongsangbuk-do, South Korea, were chosen as research subjects owing to their proximity to Jillyang-eup of Gyeongsain-si, where rural and industrial areas are in contact and urbanization is taking place for upgrading non urban area using big data. The study focuses on improving settlement conditions by upgrading the use of the current general residential area. Using big data The studied area has seen a steady increase in external influent population owing to recent urban development projects, creation of an industrial complex, and district unit planning projects. The possible expansion of influent population was reviewed, along with regional vitalization and residential environment, by projecting outcomes from reclassifying parts of the general residential area based on the big data from the 2020 and 2030 in future Gyeongsan-si basic plans. The results showed that additional residential land must be secured to accommodate the projected population growth. Therefore, this study suggests changing the use of some areas of Hwang Ze and Danggok districts, particularly to subdivide general residential areas for better district planning. Using ICT technolcgies.

Keywords Big data · Clouding · GPS (Global positioning System) · GIS tool

1 Introduction

The current legal system covering district unit planning has limitations in responding appropriately to current identified problems with the district units, particularly by reflecting the characteristics of newly planned district unit areas [1].

The research area of Hwang Ze and Danggok-li in Jillyang-eup, Gyeongsan-si, Gyeongsangbuk-do, is geographically cut-off from X by the Gyeongbu Expressway

Y. P. Geum (✉)
Catholic University of Daegu, Gyeongsan, South Korea
e-mail: kyop64@naver.com

R. Lee (ed.), *Software Engineering Research, Management and Applications*, Studies in Computational Intelligence 789, https://doi.org/10.1007/978-3-319-98881-8_13

but is located near the micro-living zones of North Jillyang and South Jillyang, as well as Jillyang industrial complex in the northwest. Hence, this region has a higher urbanization rate compared with the North Jillyang micro-living zone, with its proximity to a straight track leading to a nearby industrial complex being constructed being a significant factor. In terms of the regional background of the study area, the Middle Jillyang living zone has exceeded its planned population compared with other living zones, whereas its administrative district area is small and characterized by high urbanization owing to the high rate of industrial land use with the construction of the industrial complex. As such, most parts of the South Jillyang micro-life zone are already developed and have limitations in expanding the urbanized area to secure residential land. The current area was thus selected owing to the need for examination of undeveloped lands within a city.

The current work will examine the flows of geographical, social, and urban planning population within the Middle Jillyang living zone. Moreover, it will conduct an overall situation analysis, including problem identification regarding district unit, upgrading of use area within the existing residential area by considering the settlement conditions of expected residents, and promotion of medium- to high-density development. The upgrade of use of the area, achieved by changing the district unit plan, must be made with consideration for minimizing potential damage to neighboring residential areas and the best interest of the natural environment and cityscape. Regarding the increased income of the business operators by the upgrade, a reasonable public contribution scheme must be provided in cooperation with the corresponding city and county.

2 Conditions, Type Change, and District Unit Plan for the Target Area

2.1 Conditions of the Target Area

In 2009, parts of the 2020 Gyeongsan basic urban plan were modified. The target area was designated as a district unit in April 2012. The district unit plan was established in January 2015 and finally decided in October 2017 through the 2030 Gyeongsan basic urban plan. The major policies in the district unit plans for the target area include eco-friendly land use, effective urban redevelopment, and planned development, aimed at healthy urban development. Although the district unit plan was established in January 2015, the area has been left as a deprived area without project commencement owing to the lack of development condition.

2.2 Upgrading Land Use

Deciding the use of a unit area must be indicated by the established district unit plan. After subdivision, units are classified into classes 1, 2, and 3 zones. The classification of district unit areas can be set as concepts based on the unique characteristics of the target area and room for development. Each designated district applies different building coverage and floor area ratio values for construction. For instance, the number of floors an apartment is allowed to have within a district area is pre-determined according to the land use of unit areas.

2.3 District Unit Plan

A prior district unit plan is established on the national territory by city and *gun* (county unit) as the target area for planning to rationalize land use and to secure function, aesthetic view, and healthy environment. The district unit plan has been newly integrated into the urban planning system according to the detailed plan by the Urban Planning Act and Building Codes. Among them, class 2 district unit plan is a new policy introduced by integrating the National Land Use and Management Act and Urban Planning Act into the National Land Planning and Utilization Act, to resolve the problems of thoughtless development and ensure systematic management [2]. The National Land Planning and Utilization Act was revised to accommodate new methods of urban development. This standard distinguished between class 1 and 2 district unit plans; however, such distinction has been abolished owing to its low utility ([Implement 2012.4.15] [Act No. 10599, 2011.4.14., some parts revised]) [3].

2.4 Review on Gyeongsan-Si District Unit Plan Types

The target area is a typical non-urban area encompassing Jillyang industrial complexes I, II, and III, and the old Jaseng factory district and rural area. The area is shown in Fig. 1. The 2030 basic urban plan of Gyeongsan is focused on four parts:

(1) land use plan
(2) traffic plan
(3) population allocation plan
(4) balanced regional development.

They are currently carried out in phases with consideration for unique regional characteristics. As shown in Fig. 2, Gyeongsan's urban area spans 110.92 km^2 out of the total area of 411.78 km^2. Of the urban area, the natural green area takes up the most with 76.3% (x km^2), whereas the housing land area is 15.0% (x km^2), commercial area is 1.4% (x km^2), and industrial area is 5.2% (5.78 km^2). The district area is a mix of industrial, commercial, and residential areas.

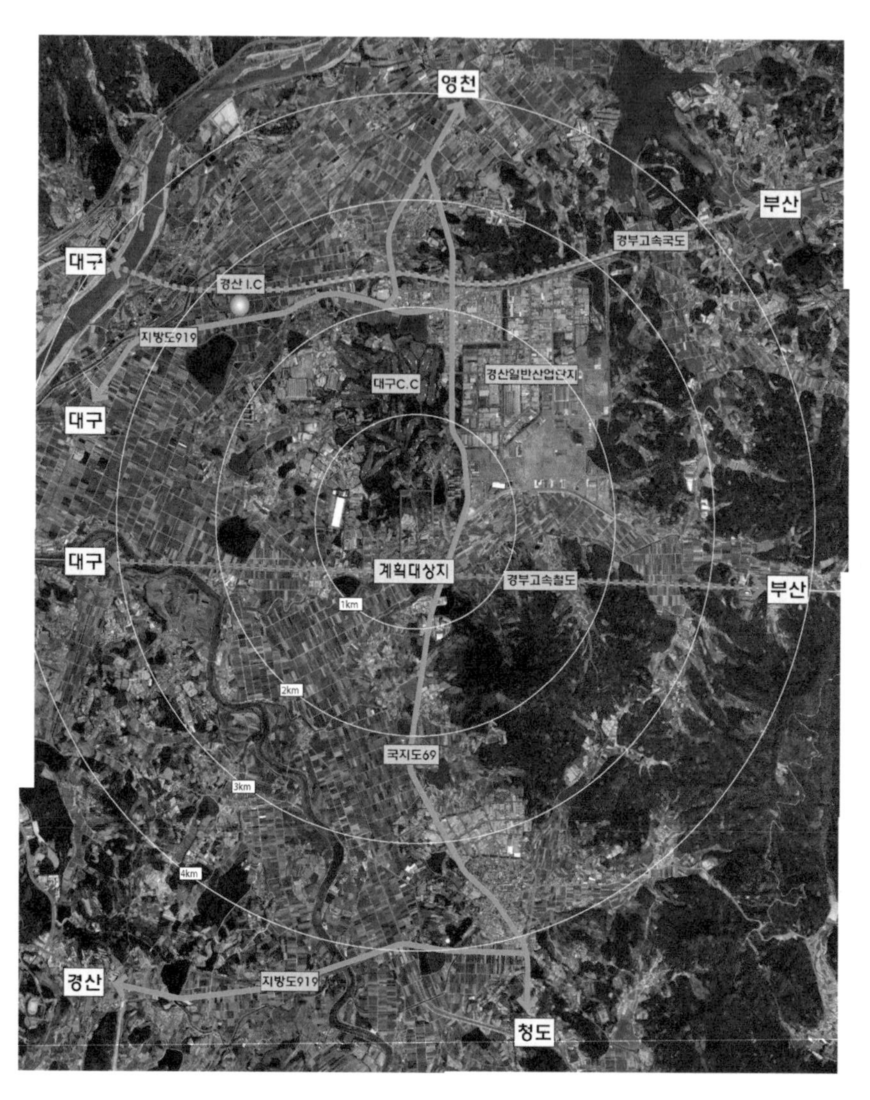

Fig. 1 Location status map of target areas

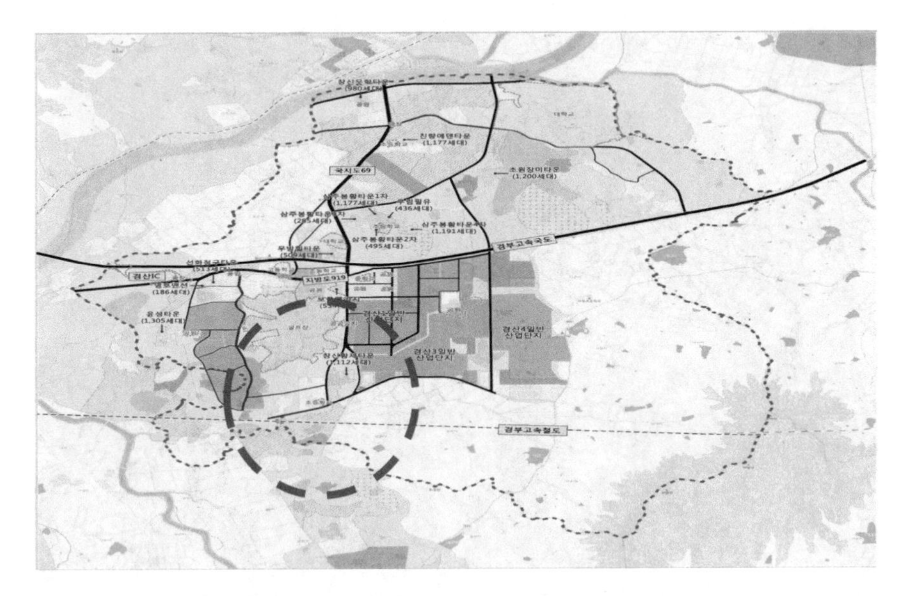

Fig. 2 Land use map

2.5 *Population Allocation Plan by Living Zone*

The population forecast material from the 2020 Gyeongsan basic urban plan shows that as of 2015, the population was 295,000 (natural increase, 264,900; social increase, 42,590), and the forecasted population at 2020 will be 400,000 (natural increase, 270,234; social increase, 130,887). With this assumption, the basic plan of the city was already established [4].

Further analysis showed that the area is located at the center of the region as shown in Fig. 3. By 2030, the Middle Jillyang life zone will have a population allocation of approximately 79,000, similar to that of the Middle Hayang life zone with 80,000 people. However, the administrative district area of Middle Hayang is roughly two times larger than that of Middle Jillyang. Figure 3 and Table 1 demonstrate that most central life zones are located near Jillyang-eup.

2.6 *Status of Urbanization Plan*

To analyze the status of the urbanization plan, the population by life zones should be examined first. In 2012, the Middle Jillyang life zone had a population of 40,526: lower compared with the Middle Gyeongsan life zone, but higher compared the Middle Hayang and Middle Jayin life zones. Middle Jillyang's administrative district area is the smallest; its 19.3% urbanization rate is relatively high. As shown in Table 2, the urbanized area of the Middle Jillyang life zone is mostly industrialized (63%);

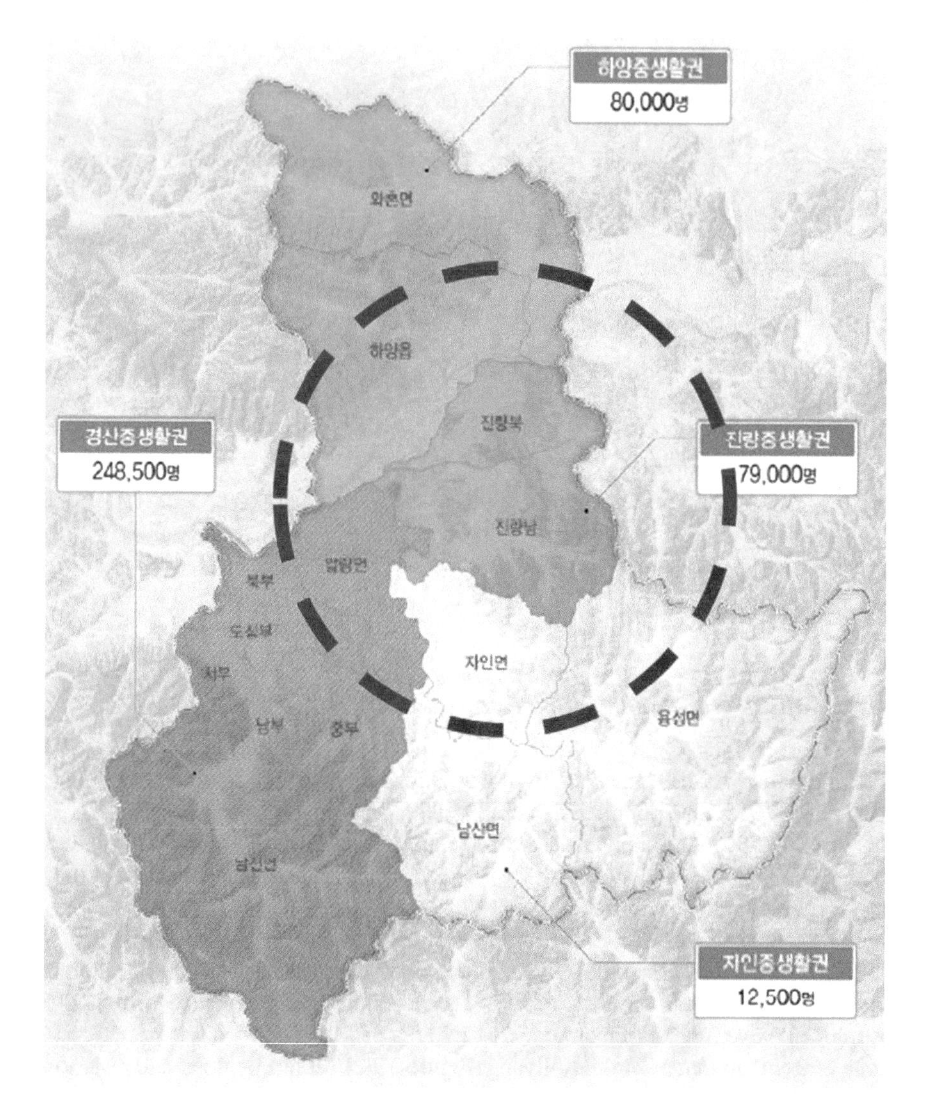

Fig. 3 2030 Basic plan design (by distribution of population)

only a third (34%) of it is residential. The population density is relatively low (135.09 person/ha) and is considered as medium-scale density (Fig. 4).

Table 1 Population change in Gyeongsan

Classification	2012	2015	2020	2025	2030
Total	252,818	270,000	320,000	370,000	420,000
Middle Gyeongsan life zone	163,517	181,000	211,500	224,000	248,500
Natural Increase	-	165,047	168,751	170,897	171,528
Social Increase	-	16,067	42,861	53,383	77,214
Middle Hayang life zone	34,298	34,000	37,500	59,500	80,000
Natural Increase	-	34,093	34,232	34,169	33,966
Social Increase	-	-	15,749	31.000	37.089
Middle Jillyang life zone	40,526	41,000	57,000	73,000	79,000
Natural Increase	-	40,818	41,482	41,992	42,516
Social Increase	-	-	15,749	31.000	37.089
Middle Jayin life zone	14,477	14,000	14,000	13,500	12,500
Natural Increase	-	13,975	13,534	12,969	12,310
Social Increase	-	-	560	560	560

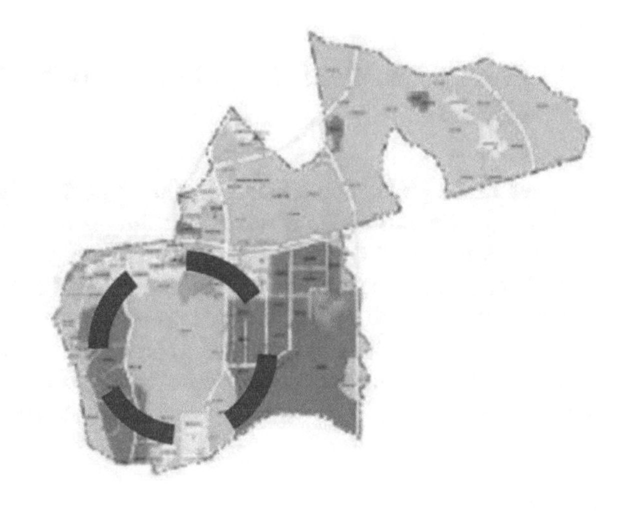

Fig. 4 2030 Basic plan design (by population distribution)

Table 2 Urbanization area and population density by life zones

Classification	Area (km²)				As of 2012 (Population) [b]	Housing Land (km²) [C]	Gross density (No. of people/ha) [B/A]	Net Density (No. of people/ha)[B/A]
	Total [A]	Urbani Zationarea	Rate of urbanization	Non Urbanization Area				
Total	411.78	29,23	7,1	382,55	252,818	17,74	6,14	142,51
Middle Gyeongsan Life zone	129.4	11,85	9,2	117,58	163,517	9,97	12,63	164,01
Middle Hayang Life zone	96,54	6,37	6,6	90,17	34,298	3,18	3,55	107,86
Middle Jillyang Life zone	46,17	8,89	19,3	37,82	40,526	3,00	8,78	135,09
Middle Jayin Life zone	139,63	2,12	1,5	137,51	14,477	1,59	1,04	91,05

(1) The urbanization area refers to the residential,commercial, and industrial areas of the urban plan (Gyeongsan use area as of December 31, 2014)
(2) Population density (population/area administrative district), including foreigners

Table 3 Population changes in the 2020 basic urban plan (*Unit* Number of persons)

Classification	2005	2010	2015	2020
Middle Jillyang life zone	43,500	50,600	59,400	65,000

2.7 *Changes in Population Allocation in the Middle Jillyang Life Zone*

Changes in the population allocation of the Middle Jillyang life zone are shown in Tables 3, 4 and 5. Table 6 shows that the population increase induced by the building of industrial complexes grew from 65,000 people in the 2020 basic plan to 79,000 people in the 2030 basic plan. The external influent population caused by new projects, such as city development projects, industrial complex establishment, and district unit planning, is estimated to be 32,376 people in the 2030 basic plan. Indeed, the external influent population increase is expected to be substantial upon the construction completion of the Gyeongsan 4 general industrial complex (Table 7).

Table 4 Population changes in the 2030 basic urban plan (*Unit* Number of person)

Classification	Current (2012)	2015	2020	2025	2030
Middle Jillyang life zone	40,526	41,000	57,000	73.000	79.000

Table 5 Induced population by urban development projects (social increase)

Classification	Living Zone	Area (1,000km2)	Induced population (No. of people)	External migration rate (%)	Influent population (No. of people)	Level	note
Jillyang Sunhwa	Jillyang	131	1,890	66,1	1,249	2nd	New

Table 6 Induced population by the construction of industrial complexes

Classification	Living zone	Area (1,000km2)	Original unit (No.of people/1,000 km2)	No. of employees (No. of people)	Extermal migrat ion rate (%)
Industrial Complex total	-	4,098	-	19,059	-
Gyeongsan 3	Jillyang	1,497	4,65	5,963	66,1
Jillyang 3	Jillyang	97	4,65	450	66,1
Gyeongsan 4	Jillyang	2,504	4,65	11,646	66,1

Classification	Occupation rate/Sales rate	No. of influent employees (No.of people)	Marriage Rate(%)	Dependents (No.of people)	Extermalinfluent population (No.of people)	Level
Industrial Complex total	-	12,598	-	17,637	30,235	
Gyeongsan 3	100	4,603.	70	6,444	11,047	2nd
Jillyang 3	100	297	70	416	713	2nd
Gyeongsan 4	100	7,698	70	1,077	18,475	2,3,4step

Table 7 Induced population by district unit plan projects

Classification	Living Zone	Area (1,000km²)	Induced population (No. of people)	External migration rate (%)	Influent population (No. of people)	Level
Jillyang Sunhwa (Housing association)	Jillyang	27	1,350	66,1	892	2nd

Table 8 Calculation demand for residential areas by life zone in Gyeongsan

Classification	Living Zone	Area (1,000km²)	Induced population (No. of people)	External migration rate (%)	Influent population (No. of people)	Level
Jillyang Sunhwa (Housing association)	Jillyang	27	1,350	66,1	892	2nd

3 GPS Status Analysis on Big Data

3.1 Changes in Life Zone Design and Business Plan

Gyeongsan 1 and 3 general industrial complexes are already established within the Middle Jillyang life zone; the Gyeongsan 4 general industrial complex is slated for completion in 2030. The mid- to long-term plan of Gyeongsan-si includes the (1) establishment of the Gyeongsan 4 general industrial complex, (2) construction of a leisure town (Muncheon area), (3) creation of residential hinterland area for industrial complexes, (4) restriction of thoughtless factory development and renewal of old industrial complex, and (5) advancement of the urban development plan of Jillyang Sunhwa district. If these plans are executed as planned, the induced population will far exceed the number already expected with respect to the existing industrial complex development. Therefore, the construction of residential hinterland areas near industrial complexes is necessary [5].

3.2 Calculating Demand for Residential Area

The residential area calculated to accommodate the population of 79,000 people planned by the 2030 Middle Jillyang life zone is set as 7.91 km^2. As the current residential area of the Middle Jillyang life zone is only 3 km^2, securing additional residential areas is urgently needed if the 2030 planned population is to be accommodated. The residential area of the Middle Jillyang life zone consists of 1.8 km^2 (60%) class 1, 0.8 km^2 (26.7%) class 2, and 0.4 km^2 (13.3%) class 3 general residential areas. As shown in Tables 8, 9 and 10, the population density within residential areas needs to be adjusted.

Table 9 Calculating residential areas by population density

Living Zone Classification	Applied population (No.of people)	Population on covered by residential area (B) (No.of people)	Density classification	Population on density (D) (No.of people/ha)	Population on distriburion (C) (%)	Population on coverage (No, of people)	Required area (A)(km²)
Jillyang	79,605	71,645	High Density	200	30	21,493	1,07
			Mid Density	150	30	21,493	1,43
			Low Density	90	40	28,658	3,18
			Subtotal	-	-	71,645	5,68

Table 10 Calculating residential areas by site area per house type

Classification		Unit	Gyeongsan	Hayang	Jillyang	Jayin
Detached House	Composition rate	%	30	40	35	50
	No. of houses	호	37,326	16,154	14,020	3,238
	Land area	km²	19,03	8,93	7,23	1,71
Row house	Composition rate	%	20	25	20	20
	No. of houses	호	24,884	10,096	8,011	1,295
	Land area	km²	3,58	1,45	1,24	0,17
Apartment house	Composition rate	%	50	35	45	30
	No. of houses	호	62,209	14,135	18,025	1,943
	Land area	km²	5,42	1,24	1,63	0,16
Total		km²	28,02	11,9	10,15	2,04

3.3 *Analysis of Developable Area*

Urban–rural integrated cities were born to induce balanced development between the urban and rural areas. However, imprudent developments that ignore regional characteristics have caused even greater regional imbalance [6]. To utilize the existing land use plans efficiently, the developable areas were compared, evaluated, and sorted by condition, as shown in Fig. 6. The framework of the analysis criteria is shown in Table 11, whereas the process of analysis on the developable area is shown in Fig. 5. The analysis sought to identify existing developed, development restricted, and development prohibited areas based on the method of using the geographic information system (GIS) tool. Figure 7 displays the final result of the comparative analysis of developable areas by condition: 126.25 km^2, or 30.7% of the total area of 411.78 km^2, is analyzed as developable area, whereas 156.60 km^2, or 38.3%, is considered as undevelopable area. Meanwhile, the development control area accounts

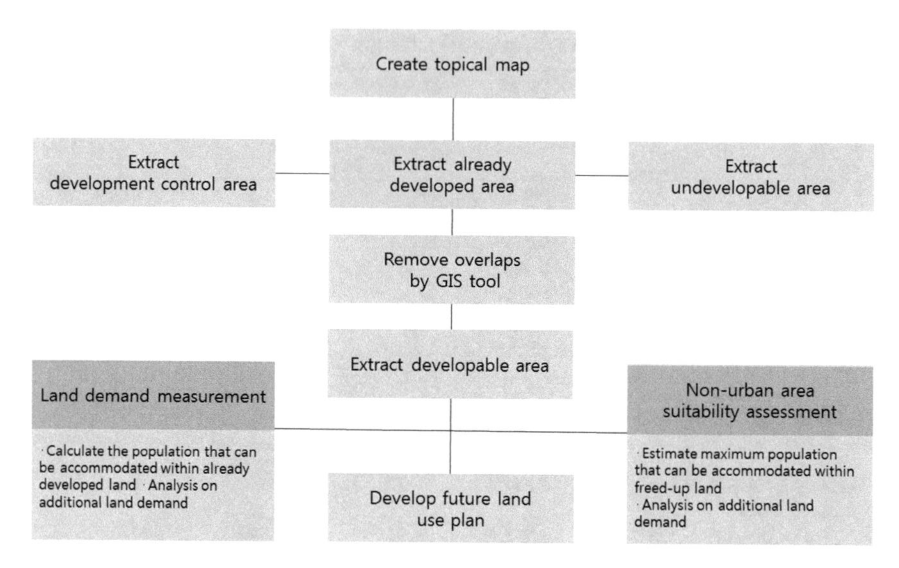

Fig. 5 Process of analysis of developable area

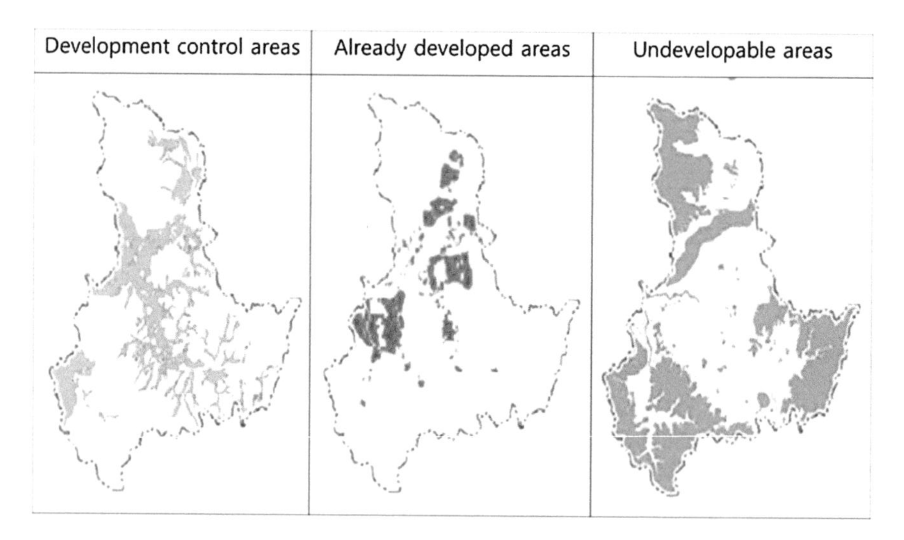

Fig. 6 Comparative analysis of developable areas by condition

for 21.5% (88.69 km^2) of the total area. Already, 39.24 km^2, or 9.5% of the total area, is considered as already developed.

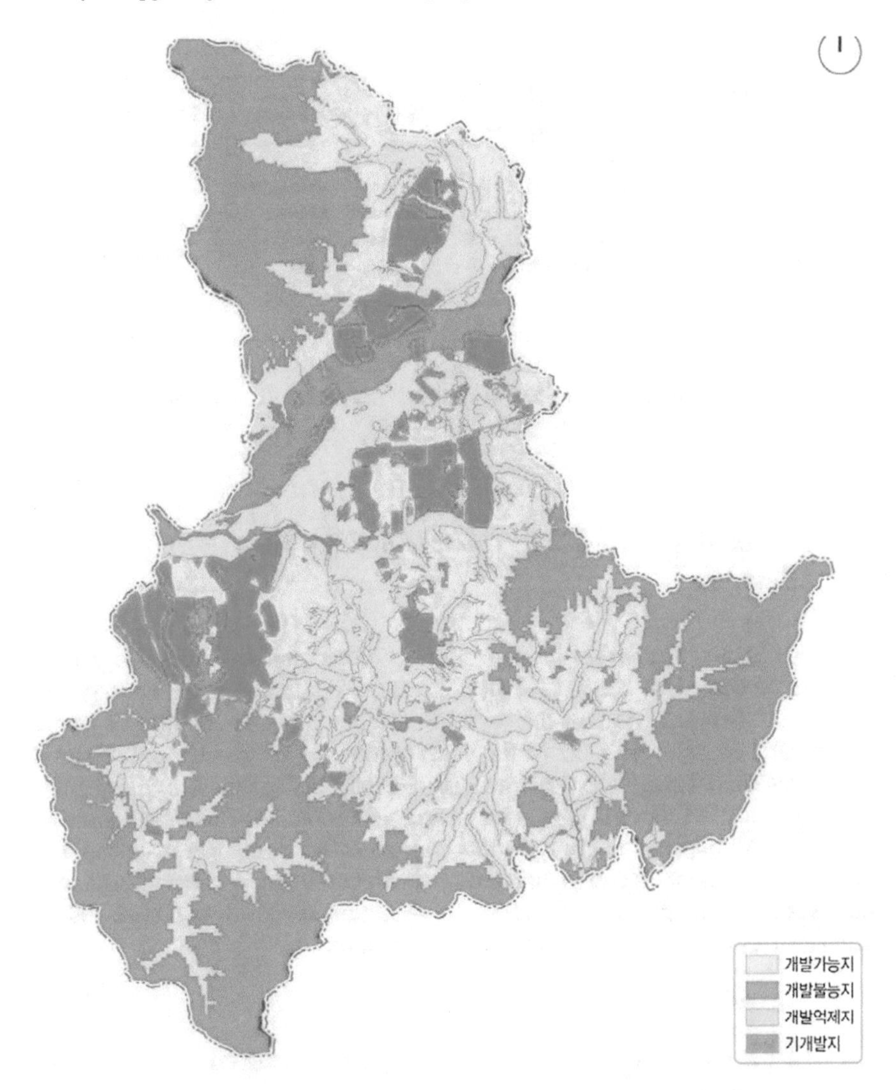

Fig. 7 Results of comparative analysis of developable areas by condition

3.4 Analysis of Undeveloped Area

In the urbanized area, the undeveloped area consists of land with categories of *jeon* (field), *dap* (rice field), and *yim* (forest). Out of 9.22 km^2 of urban area in the Middle Jillyang life zone, only 0.18 km^2 is undeveloped, which is insufficient considering the induced population. As the Middle Jillyang life zone has a small administrative area and a large urbanization rate, it would be necessary to expand the non-urban area

Table 11 Framework of analysis criteria [7]

Development control area	Already developed area	Undevelopable area	Developable area
Land where development is deferred or restricted	Commercial, residential, industrial areas in the city management plan; currently developing area and areas with future development plans	Area where development is not possible physically, naturally, and environmentally	Available land after excluding already developed, undevelopable, and development control areas

by evaluating the undeveloped land within the city when drawing the management plan.

3.5 *Status of Space Facility Change by Unit Plan (Type) Change*

As shown in Fig. 8, changes in the confirmed district unit area are analyzed and found to be desirable. The area of class 1 general residential area in change plan I should be reduced. Parts of the south section should be set as class 1 general residential area, whereas parts of the north section should be changed to class 2 general residential area, as shown in change plan II about Big Data and GPS analytics [8]. The allocation plan and population inducement should consider the skyline and the region. Modifying the use type of certain parts in this manner may lead to an increase of the floor area ratio and easing of height. Hence, the permitting agency should recommend plans that correlate the allocation of apartment houses with the neighboring topographical and surrounding conditions.

3.6 *Changing Use Area and Management Direction for Urban Use Areas*

This study evaluated various aspects of the 2030 basic urban plan, which encompasses the subordinate life zone and population allocation plans and residential demand estimation, namely, excess or deficiency in housing land and need for additional housing land given the inflow of external population, for which preparations are needed. The need for changes of use in district units was found to be necessary after the evaluation. Hence, the final results were derived from the analysis of different aspects. The results are as follows.

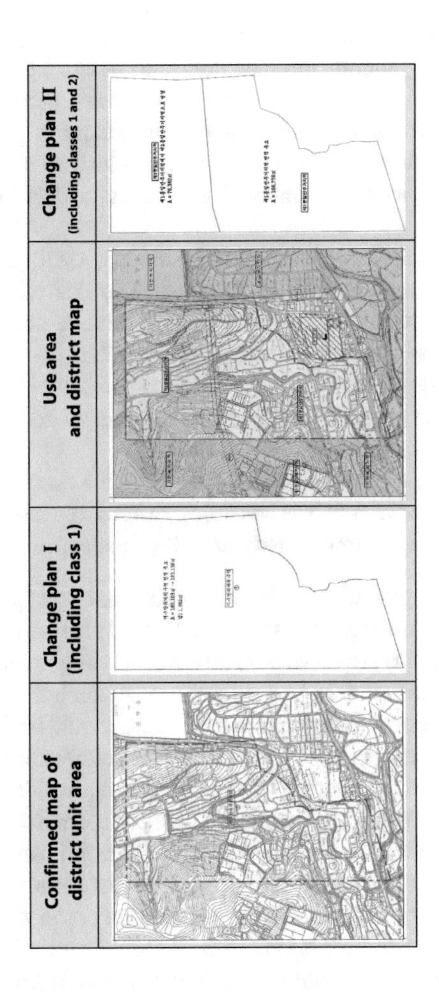

Fig. 8 Change plan

(1) Various elements must be examined synthetically before formulating the decision for changing land use. Once decided, adjustments need to coincide with the enactment and revision of legislation as well as related plans.
(2) If area use change is needed to accomplish an urban management plan such as the basic urban plan, then such change in the district unit plan, urban planning projects, and redevelopment projects should be within the related plan; changes should be characterized by total planning, as in the case of Gyeongsan-si in this work, which considers CBD and life zone planning.
(3) The applicable city should refer to policy to assess the reorganization of a city management plan and adjustment of use areas (in detail) for the efficient use of land, even at the minimum range, from the priority area. The city must consider the fact that land use changes rapidly with the completion or on-going development of projects in neighboring areas as well as the establishment of influential areas. [9].
(4) Cities should adopt the basic principle of evaluating on the assumption of establishing detailed plans for changing land use. The change should result in reasonable public contribution, by principle and criteria, and the incomes of the newly profited people should contribute to the public good.
(5) Through regional-level pre-planning, like life zone planning, to improve quality of infrastructure, an efficient supply system closely related to everyday life may be established.
(6) As direction for regional management on urban area usage, the current land use should be maintained as much as possible to protect the residential environment exclusive to low-height, low-density detached housing.

The basic principle of segmenting residential area classes for managing the use and density suitable for exclusive residential areas should also be maintained. The city and *gun* should arrange ways to make adjustments according to the above. Moreover, the fixed range of use area should be set to minimize the influence on surrounding areas. Extreme differences with surrounding use areas should be avoided. The core of such plan is the people-first principle. Therefore, the suitable residential density and height standard should be considered. Homogeneity of the residential area should not be lost by bringing in uses that are against the residential environment. Upgrading of a residential area class must be performed with minimized damage on surrounding residential areas, such as solar access rights and privacy. Considering the surrounding natural environment, cityscape, and change in condition, semi-residential areas must be managed as combined-use areas centered on the influential area to complement urban-type residence as well as commercial and business functions.

4 Conclusion

This study compared the Middle Jillyang life zone with the other life zones in Gyeongsan in terms of population, urban administrative area, population density, and urbanization rate. The land use areas were also determined and analyzed. With reference to the 2020 and 2030 basic plans, the study found a significant increase in the population plan for the Middle Jillyang life zone, attributed to urban development projects, industrial complex establishment, and district unit planning projects. Indeed, the inflow of external population will only increase further, requiring the construction of residential areas around industrial complexes. The residential area demand analysis revealed a large gap between the ideal (7.91 km^2) and actual (3 km^2) area to accommodate the 2030 planned population of 79,000 for the Middle Jillyang life zone. Additional land for residence may be secured by expanding the residential area or changing the density within the residential area. However, as the Middle Jillyang life zone has a small administrative area and high urbanization rate, which limit expansion, undeveloped land within the city should be evaluated as much as possible, while also avoiding the expansion of non-urban areas. The population and GIS tool analyses showed that density adjustment is needed through un-zoning within the existing residential areas. The city and *gun* will need using big data and GPS [10] to undertake use upgrade by changing district units for the sake of the projected increased population by inflow.

Acknowledgement This work was supported by research grants from Daegu Catholic University in 2018.

References

1. Yeo, S., Kang, G.: A study on the typification of district planning area considering regional characteristics. Korea Spat. Plann. Rev. 151–162 (12 pp.) (2015)
2. Bae, Y.K.: Study of Land Method. Sejong Publishing Company, pp. 1–23 (23 pp.) (2010)
3. Ministry of Land, Infrastructure, and Transport: Terminology Dictionary on Land Use. Jinhan M&B (2016)
4. Gyeongsan-si: Basic urban planning for the Gyeongsan City in 2020. http://gbgs.go.kr/main.jsp (2009)
5. Hun, P.G., Gwan, J.S., Yeong, C.W.: Urban Regeneration and Living Environment Reformation Project, Ministry of Land, Infrastructure and Transport Coverage, vol. 3, pp. 9–12 (2015)
6. Hyun, P.G., Kwan, J.S., Meong, C.W.: Regional characteristics analysis for sustainable development in urban and agricultural integration. J. Korean Assoc. Geographic Inf. Stud. **3**(2), 37–47 (2000)
7. Changwon-si: Basic urban planning for the changwon, land use plan 'How to check the paper to be Urbanization'. https://www.changwon.go.kr/main.do (2015)
8. Kim, K.H.: A Study on the System and Cases of Urban Regeneration in Foreign Countries. Urban Regeneration Project (2011)
9. Geum, Y.P.: A Study on the Urban Regeneration Plan for Sustainable Creative Cities, p. 102. Graduate School of Gyeongil University (2016)
10. Yang, J.M.: Handling GPS Location Data in Criminal Investigation, pp. 225–247. Gyeongsang National University Law Institute (2016)

Simulation of Flood Water Level Early Warning System Using Combination Forecasting Model

Kristine Bernadette Barrameda, Sang Hoon Lee and Su-Yeon Kim

Abstract This research explores the use of BPNN and SVM techniques as a combined model using the Minimum Variance (MV) method to predict the upcoming flood water level events in Calinog River, Iloilo, Philippines. Rainfall and water level values are utilized as predictive variables to evaluate the performances of the individual models and the proposed combined-model as applied in the datasets. Root Mean Squared Error (RMSE) is used as a performance indicator of the trained models. Various simulation experiments are conducted to investigate the performance of the proposed model and the results show that the proposed combined-model of BPNN and SVM with their identified best control parameter values, produced a good predictive result as compared to the individual performances of SVM and the BPNN model. The proposed model yields better results that will surely help improve the effectiveness of the implementation of plans and policies of the disaster risk management of the local government unit and Iloilo Province as a whole.

Keywords Support Vectors Machine · Back Propagation Neural Networks
Combined forecasting · Prediction modeling · Flooding

1 Introduction

In the late 90s, Bates and Granger first introduced the idea of a combination model for forecasting to address an issue on the selection of the best individual model to solve a problem, since individual models have different kinds of approach in processing

K. B. Barrameda · S. H. Lee · S.-Y. Kim (✉)
School of Computer and Information Engineering, Daegu University, Gyeongsan 38453, Republic of Korea
e-mail: sykim@daegu.ac.kr

K. B. Barrameda
e-mail: kbbarrameda12@gmail.com

S. H. Lee
e-mail: prolee@prolee.net

R. Lee (ed.), *Software Engineering Research, Management and Applications*, Studies in Computational Intelligence 789, https://doi.org/10.1007/978-3-319-98881-8_14

variables, where they mentioned that (a) one model based its processes to different variables or information that other model/s does not consider and; (b) The other model differentiates with making assumptions about the structure of relationships between various variables [1]. The overview of several methods or concepts on how to effectively generalize models and acquire best accuracy of combination prediction models are presented by [2, 3]. Prediction combination models have a wide range of application in the following areas mentioned in [2] such as Meteorological Data, Gross National Product (GDP), Inflation, Interest Rates and Money Supply, Stock Returns, Census Data etc. This study is focused on building a river flood water level monitoring and prediction framework with warning system integration through the application of a combination forecasting model. SVM and ANN methods are currently one of the widely-used techniques in time-series analysis, particularly in predictive modeling as well as in some other applications like text mining, financial and energy where the performance of the two algorithms are seen to be very high and effective. Therefore, we would like to explore the application of these methods as a combined-forecasting model for flood water level prediction that would improve the accuracy of the result in predicting flood occurrences in Calinog, Iloilo Province.

2 Methodology

2.1 Study Area and Data

The study area is the Calinog river at the Jalaur river basin in Iloilo Province, Philippines. The town of Calinog is located in the central part of Panay Island. It is located between geographical coordinates X 455017.7096 Y 1241214.0575 northeast and X 418450.7932 Y 1225637.0838 southwest using UTM 51 Luzon Datum. The urban areas are mostly located at the lowland part of the town. One of the issues in the downstream area is the need to reinforce with the Municipal Disaster Risk Reduction Management committee on rainfall monitoring, stream flow monitoring and its relation to disaster management and mitigation [4]. The data used to train the proposed model are extracted 3-year period data from 01/01/2014 to 01/01/2017 of the automatic weather gauging station at Calinor River managed by the Department of Science and Technology Region VI. Flood water levels can be forecasted based on (a) rainfall data; (b) previous water levels; and (c) a combination of both dataset [5]. Therefore, we utilized the following independent variables such as dateTimeReceived, rain_value and waterLevel from the specified gauging station.

2.2 Systems Design and Architecture

ANN and SVM algorithms are very effective in solving non-linear problems and therefore, we apply these methods in our study through a combination forecasting model using weighted average technique, the Minimum Variance (MV) method. The flood water level prediction result, which is the final prediction output, will be loaded and utilized for analysis in the integrated warning system. Data collection, data analysis, data warning response recommendation and data dissemination are performed as shown in Fig. 1.

(A) Back Propagation Neural Networks

A clear definition of BPNN algorithm is presented in [6]. Applied in this study is the basic structure of BPNN, which is one input layer, one hidden layer and one output layer. During training, the size of the tasks taken in the weight space is a component of various internal network parameters including the values of the learning rate, momentum, error function, epoch size and the transfer function.

(B) Support Vectors Machine

The separation of two factors in a dataset in real-world scenarios is not generally easy but SVM deals with this issue by (1) allowing a couple of data to the wrong side of the hyperplane by displaying a predetermined parameter C that demonstrates the trade-off between the maximization of margin and misclassifications and; (2) using kernel functions like polynomial, sigmoid, and linear and Radial Basis Functions (RBF) to add more dimensions to the low dimensional space, that two classes could be divisible in the high dimensional space [7]. In this study, a simple SVM method with kernel function is implemented. The Kernel function and C (penalty parameter

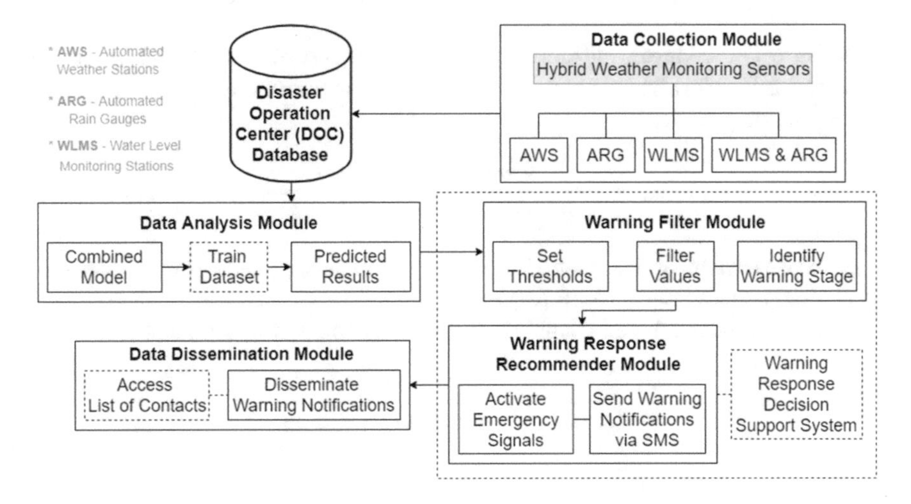

Fig. 1 System framework overview

of the error) are the SVM parameters that users could play with (trial and error) to effectively improve the model performance.

(C) The Proposed Combination Model

The individual prediction models are BPNN and SVM where BPNN prediction results is P1; SVM prediction results is P2 and the weighted combined prediction is Pc which is defined as:

$$P_c = w_1 P_1 + w_2 P_2 \quad (1)$$

where w1 and w2 are respective weights of P1 and P2. The summation of wi = 1. The main reason that the sum of the weights is constrained to add up to one is to ensure unbiasedness [8]. Prediction errors are e1, e2 and ec defined as, e1 = y − P1, e2 = y − p2 here y is the actual values, and P1 and P2 are two individual prediction results with their prediction errors and;

$$e_c = w_1 e_1 + w_2 e_2 \quad (2)$$

2.3 *Determination of the Combination Optimal Weights*

In this study, Minimum Variance (MV) method is applied to minimize the error variance of the combination model for prediction. To find the optimal weights for each of the models the following equations are defined, as adopted from [9] and as described in [10]. Combined error variance is defined as:

$$Var(e_c) = w_1^2 Var(e_1) + 2w_2 w_2 Cov(e_1 e_2) + w_2^2 Var(e_2) \quad (3)$$

where Var(e1) and Var(e2) are individual error variances of the models and $Cov(e_1 e_2)$ for covariance. Thus, to calculate the minimum error variance for w_1, weight is set to the following:

$$w_1 = \frac{Var(e_1) - Cov(e_1 e_2)}{Var(e_1) + Var(e_2) - 2Cov(e_1 e_2)} \quad (4)$$

Since we have separately trained the individual models, P_1 and P_2 are independent prediction models, hence, $Cov(e_1 e_2) = 0$.

Therefore, the combined weight coefficients, w_1 and w_2 are defined as:

$$w_1 = \frac{Var(e_2)}{Var(e_1) + Var(e_2)}, w_2 = \frac{Var(e_1)}{Var(e_1) + Var(e_2)} \sqrt{w_1^2 + w_2^2} = 1 \quad (5)$$

2.4 Experimental Plan and Prediction Evaluation Criterion

The main steps of the experiments are based on the combined-method system model, illustrated in Fig. 2. The collected data are preprocessed and trained separately using both models. The two prediction results are weighted and combined to get the final prediction. In this study, to verify the efficiency of the proposed combination model, various experiments are implemented as discussed in A, B, C, D & E:

(A) Experiment 1: Estimate SVM Parameters

In this experiment, we identify the best parameters on the development set. Scikit-Learn SVM library incorporates kernel functions, for example, linear, polynomial, sigmoid and Radial Basis Function. To train SVM models, the user must choose the correct C parameter value and any required kernel parameters which are presented in [11] shown in Table 1, where the Kernel type and minimum and maximum values of C are given.

(B) Experiment 2: Estimate BPNN Parameters

This experiment applies the four parameters such as the number of hidden layer nodes, learning rate, momentum, and epoch in the ANN learning process as the design of experiment factors as cited in [12], where every factor has 2 levels of set values as shown in Table 2.

(C) Experiment 3: Compare Combination Weight Techniques

In this experiment, two widely known combining techniques, (1) Equal Weights (EW) and; (2) Minimum Variance (MV) methods are implemented and compared. The goal of this experiment is to compare the two different results against each other

Table 1 SVM control factors given values

Control factors	Type	Minimum	Maximum
Kernel	Linear, polynomial, sigmoid, RBF	–	–
C	–	1	1000

Table 2 BPNN level settings for each parameter

Control parameters	Level	
	Minimum	Maximum
1. Number of hidden layer nodes	8	20
2. Learning rate	0.3	0.9
3. Momentum	0.5	0.9
4. Number of epochs	20,000	40,000

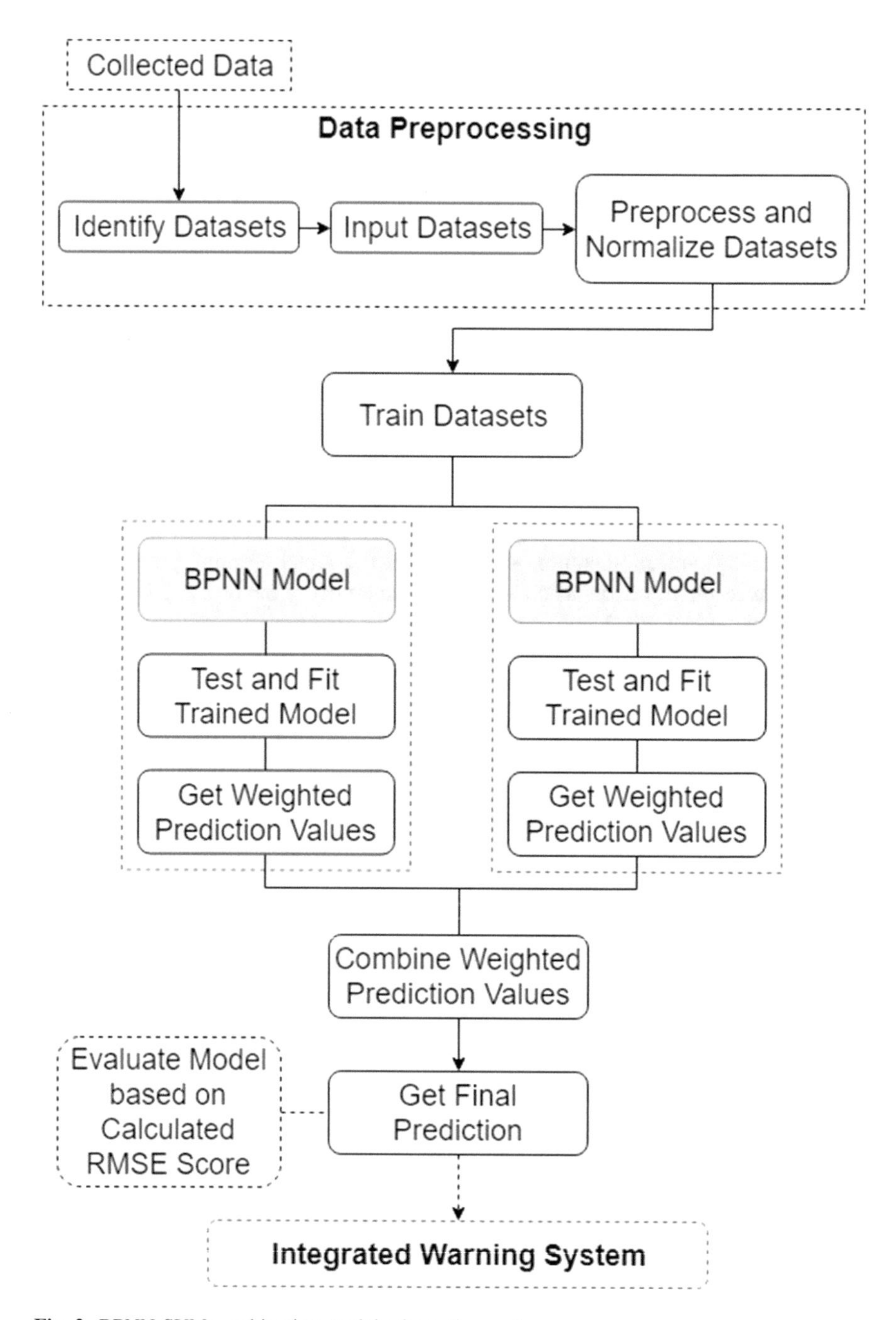

Fig. 2 BPNN-SVM combination model schematic graph

and to evaluate the performance of our proposed model with MV application as presented in Eq. (1). EW is defined as:

$$w_i = \frac{1}{N} \quad (6)$$

where (i = 1, 2, 3…N) and N is the number of forecast models.

(D) Experiment 4: Utilize Peak Values as Dataset

Dataset values during rainy and stormy season are utilized in this experiment as training set which could include flood events.

(E) Experiment 5: Use Multiple Number of Test Sets

In this experiment, various number of test sets is used to evaluate the performance of the models, which will be compared based on the performance indicator, the RMSE scores. The prediction performances of the models are evaluated through calculating the RMSE score defined as:

$$RMSE = \sqrt{\frac{1}{n}\sum_{i=1}^{n}(y_i - P_i)^2} \quad (7)$$

where $y_i(i = 1, 2 \ldots n)$ are the actual values, $P_i(i = 1, 2 \ldots n)$ as the prediction results and n represents the number of evaluation periods.

2.5 *The Model Training Simulation Software*

For implementation, the simulation software that we used are Sci-kit Learn-0.15.2, Numpy-1.9.1, Scipy-0.14.0 and Pandas 0.15.1 for training and testing the models and Matplotlib-1.4.2 for data visualization. These are open-source packages implemented in Python-2.7.12.

2.6 *Integrated Warning System*

The predicted flood water level data are loaded and utilized for warning data analysis. The thresholds are set for each warning stage with automatic warning response recommender. Iloilo Province has currently a community-based early warning system which is also an effective way of giving notifications to the community about an impending flood event. However, in this study, we designed and improvised the warning system to be more reliable, accurate and cost-effective with less human intervention.

Table 3 SVM and BPNN best control parameter results

SVM control factors	Values	BPNN control factors	Values
Best kernel	Linear	Best no. of neurons	8
Best C	100	Best learning rate	0.3
Accuracy score	0.966	Best momentum	0.5
		Best epoch	20,000
		Accuracy score	0.956

Table 4 Equal weights/minimum variance technique results based on RMSE scores

Performance indicator	Models (weights w1 = 0.5, w2 = 0.5)			Models (w1 = 0.4180, w2 = 0.5675)		
	BPNN	SVM	Combined	BPNN	SVM	Combined
RMSE (2016/01–2017/01)	0.0594	0.0503	0.0316	0.0594	0.0503	0.0290
RMSE (2014/01–2017/01)	0.0505	0.0319	0.0284	0.0505	0.0319	0.0225

Table 5 Peak values with equal weights/minimum variance results based on RMSE scores

Performance indicator	Models (weights w1 = 0.5, w2 = 0.5)			Models (w1 = 0.5683, w2 = 0.4316)		
	BPNN	SVM	Combined	BPNN	SVM	Combined
RMSE (2016/01–2017/01)	0.0574	0.0392	0.0280	0.0574	0.0392	0.0271
RMSE (2014/01–2017/01)	0.0470	0.0299	0.0210	0.0470	0.0299	0.0202

Table 6 Comparison of model performances based on RMSE score with different test sets

Test sets (%)	Performance indicator	BPNN	SVM	Combined
10	RMSE	0.0610	0.0789	0.0479
20		0.0600	0.0581	0.0329
30		0.0505	0.0319	0.0225

3 Results

(A) Experiments 1 and 2 Model Parameters Estimation Results (Table 3)
(B) Experiment 3 Combination Weight Techniques Results (Table 4)
(C) Experiment 4 Peak Values as Dataset Results (Table 5)
(D) Experiment 5 Multiple Number of Test Sets Results (Table 6)

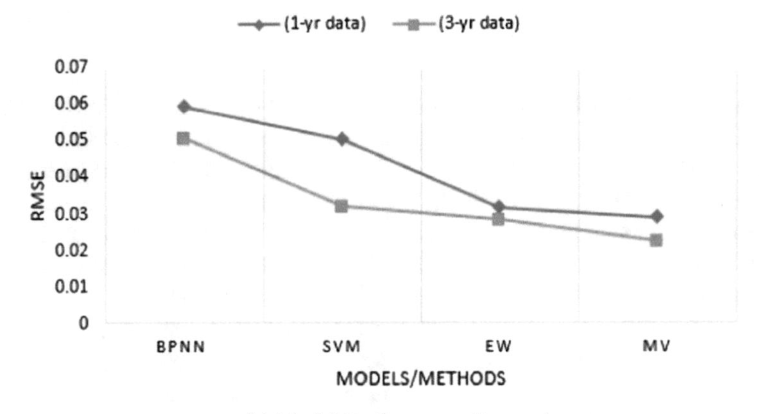

(a) Model Performance Comparison

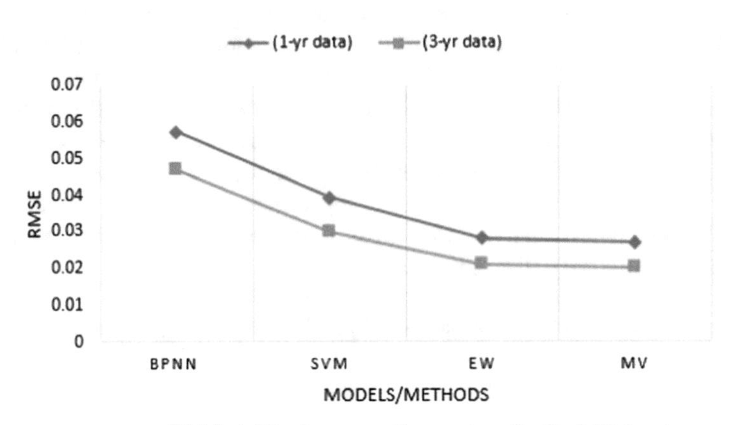

(b) Model Performance Comparison for Peak Values

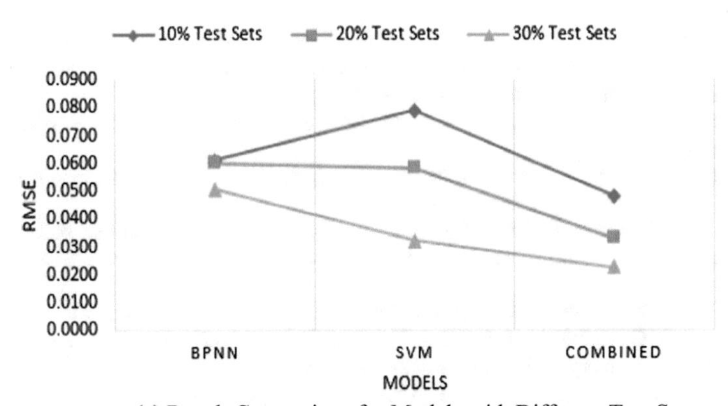

(c) Result Comparison for Models with Different Test Sets

Fig. 3 Experiment results

4 Discussion

The results display that the proposed combined model with Minimum Variance (MV) method constantly produced the best prediction results with minimum RMSE scores as presented in (a), (b) and (c) with 1- and 3-year data performance comparison.

Figure 3 implies that as the number of dataset values increases the model performances become more robust and highly accurate. We evaluated the performance results of the three models and our proposed combination model with Minimum Variance method showed the best performance. The combined model is also shown to be an effective model for flood water level prediction in Iloilo Jalaur River. Overall, the results show that daily rainfall and water level data variables could be used to accurately predict flood water level. Therefore, our approach is a reasonable method to apply for flood water level prediction, particularly if a limited number of variables are available. This study is only confined to one type of BPNN and SVM. Several other methods could also be considered to run our datasets. Further research could still be done to improve the performance of the flood water level forecasting models through examining various training algorithms. It is likewise possible to consider other input parameters such as water flow, gauging distance and river area measurements, and different forecast time frames, which may improve the predictive performance of the system. We would like to further investigate the roles of these parameters in flood water level forecasting and implement the proposed system framework in a real-world environment in the future.

Acknowledgement This research was supported by the Daegu University Research Grant.

References

1. Bates, J.M., Granger, C.W.J.: The combination of forecasts. OR Soc. **20**, 451–468 (1969)
2. Timmerman, A.: Forecast Combinations. Retrieved from http://www.oxford-man.ox.ac.uk/sites/default/files/events/combination_Sofie.pdf (2013)
3. Clemen, R.: Combining forecasts: a review and annotated bibliography. Int. J. Forecast. **5**, 559–583 (1989)
4. Municipality of Calinog: Office of Watershed Management Board. State of Calinog Watershed Report Alibunan and Upper Jalaur Rivers (2013)
5. Chang, F.J., Chen, P.A., Lu, Y.R., Huang, E., Chang, K.Y.: Real-time multi-step-ahead water level forecasting by recurrent neural networks for urban flood control. J. Hydrol. **517**, 836–846 (2014)
6. Rojas, R.: Neural Networks: A Systematic Introduction. Springer, Berlin (1996)
7. Lee, M.C., To, C.: Comparison of support vector machine and back propagation neural network in evaluating the enterprise financial distress. Int. J. Artif. Intell. Appl. **1**, 31–43 (2010)
8. Yu, W., Liu, T., Valdez, R., Gwinn, M., Khoury, M.J.: Application of support vector machine modeling for prediction of common diseases: the case of diabetes and pre-diabetes. BMC Med. Inform. Decis. Mak. **10**, 1–7 (2010)
9. Adhikari, R., Agrawal, R.K.: Combining multiple time series models through a robust weighted mechanism. In: 2012 1st International Conference on Recent Advances in Information Technology, Dhanbad, pp. 455–460 (2012)

10. Nan, X., Li, Q., Yu, J., You, Z.: Wind speed forecasting based on combination forecasting model. In: 2010 International Conference of Information Science and Management Engineering, Xi'an, pp. 185–189 (2010)
11. Jeong, D., Kim, Y.O.: Combining single-value streamflow forecasts—A review and guidelines for selecting techniques. J. Hydrol. **377**, 284–299 (2009)
12. Staelin, C.: Parameter selection for support vector machines. Hewlett-Packard Company, Tech. Rep. HPL-2002-354R1 (2003)

A Study on the Components that Make the Sound of Acceleration in the Virtual Engine of a Car

Sang-Hwi Jee, Won-Hee Lee, Hyungwoo Park and Myung-Jin Bae

Abstract Since the advent of steam-based automobiles for the first time in the 18th century, automotive technology has improved dramatically over the past 100 years. Recently, the 5GA (5G Automotive Association) was launched in time for the 5G era. For the 5Gs, eco-friendly automobiles were named. Currently, most of the automobile market uses internal combustion engines. Internal combustion engine cars have the disadvantage of causing environmental pollution and air pollution. Environmentally-friendly automobiles have been developed to be capable of high-speed driving, and are superior in noise, fuel efficiency and environmental factors compared to vehicles using internal combustion engines. However, there is a disadvantage in that a virtual engine sound is not as full as the richness of an internal combustion engine's sound during acceleration. Therefore, in this paper, we study the virtual engine sounds that can enhance the feeling of acceleration by controlling the playback speed of a virtual engine sound. Experimental results showed that an engine sound with 4 s playback time adjusted from the engine sound of 1–5 s can enhance the feeling of acceleration the most. An MOS test showed that the virtual engine sound was not much different from the engine sound of an existing engine.

Keywords Engine sound · Virtual engine sound · Acceleration · Formatting Hearing characteristic

S.-H. Jee · W.-H. Lee · H. Park (✉) · M.-J. Bae
Department of Telecommunication Engineering, Soongsil University Sori Engineering Lab
Sangdo-Dong, Dongjak-Gu, Seoul, Korea
e-mail: pphw@ssu.ac.kr

S.-H. Jee
e-mail: slayernights@ssu.ac.kr

W.-H. Lee
e-mail: wbluelovew@ssu.ac.kr

M.-J. Bae
e-mail: mjbae@ssu.ac.kr

R. Lee (ed.), *Software Engineering Research, Management and Applications*, Studies in Computational Intelligence 789, https://doi.org/10.1007/978-3-319-98881-8_15

1 Introduction

In 1770, as steam engine technology developed after the Industrial Revolution of the mid-17th century, Nicolas-Joseph Cugnot developed a steam-based automobile. Over the next 100 years, automotive technology developed [1]. Today, the world's leading companies and researchers have launched the 5G Automobile Association (5GAA) to develop and implement solutions for vehicle systems in order to better design and commercialize them. Among them, eco-friendly automobiles were classified as core industries of the next generation. In 1997, a conference was held in Kyoto, Japan to define a target of greenhouse gas reduction for advanced countries in order to prevent global warming. In the following year, the Kyoto Protocol was passed through the 3rd Conference of the Parties to the Convention on Climate Change. The main content was to reduce greenhouse gases by 2012 [1]. In order to reduce air pollution and greenhouse gases, various countries have implemented various measures based on this protocol. One of them was the development of environmentally friendly vehicles. Currently, automobiles based on internal combustion engines have the highest share of the automobile market. However, internal combustion engine automobiles have the disadvantage of causing environmental and air pollution. Among environmentally-friendly vehicles, there are electric, hydrogen, and bio-fuel cars [2]. Electric and hydrogen vehicles are now being developed with high-speed technology, which has superior noise reduction, fuel economy, and better environmental impacts compared to vehicles powered by internal combustion engines [3]. However, drivers who are aware of the speed and engine condition of an engine through the engine sound of an internal combustion engine are not able to prevent speeding or engine failure in advance due to the silence of an environmentally friendly vehicle engine [4]. Pedestrian traffic accidents can be caused by the unawareness of eco-friendly vehicles and the unawareness of these vehicles approaching. A Virtual Engine Sound System (VESS) was developed to prevent human injury and property damage caused by electric vehicle car accidents [5]. This is a system allows pedestrians to recognize a vehicle approaching by artificially generating an engine sound when a vehicle is running at low speed [5, 6]. However, the driving sound might not be efficiently transmitted depending on the acoustic and sound transmission characteristics of the speaker reproducing the sound. In addition, there is a disadvantage in that the engine sound reproduced when accelerating does not reach the driver. Drivers hear the car's engine sound due to the existing Pre-learning and listening effect.

However, if one hears a developed virtual engine sound, one will get a different feeling compared to the existing engine sound. Especially during acceleration, a driver feels pleasure through the sense of speed and the engine sound accordingly, but there is not enough with an existing virtual engine sound. In this paper, based on a virtual engine sound, we want to investigate how fast an engine sound should be generated during acceleration so that a driver could feel acceleration. Chapter 2 deals with the basic elements of sound. Chapter 3 explains existing engine sounds. Section 4 presents the results of experiments and conclusions in Sect. 5.

2 Basic Elements of Sound

2.1 Sound Generation

Sound generation is caused by periodic actions such as friction, impulse, and explosion between objects. Sounds of different frequencies are generated depending on the part from where the sound is generated and the cause. In the case of a car, the sound generated by an engine is based on an explosion sound, and the shape and size of the sound are formed according to the weight, size, and displacement of the vehicle [7].

2.2 Three Elements of Sound and Resonance

The size of a sound refers to the amount of energy that moves during a unit of time, and when the amplitude increases, the size of the sound increases. Sound has energy which propagates as the sound increases. Since an oscillating object has energy proportional to the square of the amplitude, air molecules that transmit sound also have energy proportional to the square of the amplitude of the oscillation. The energy transmitted by a sound during a unit of time, that is, the power, is proportional to the square of the amplitude multiplied by the speed of sound. The height of a sound is determined by the number of vibrations of the wave, and when the frequency is large, the height of the sound is increased. The high frequency of a sound wave is a distinction between treble and bass, and is determined by how many vibrations occur in one second. The higher the vibration, the higher the frequency, the lower the vibration, the lower the frequency. The range of frequencies a human can hear (audible frequency) is 20–20,000 vibrations per second. Even though two sounds can be of the same size and the same sound, the waveform of the sound wave can appear differently if the duration is different. This means the duration of the sound is specified by the vibration time of the sound source and is directly proportional to the vibration time of the sound emitting object [7, 8].

2.3 Ear Structure and Characteristics

Ears correspond to transducers that convert sound waves into electrical signals of the auditory nerves and are broadly divided into middle auditory and inner auditory. Figure 1 shows the structure of the ear. The outer ear plays the role of amplifying the sound through the ear canal through the throat, ankle and spine. The middle ear is the way to a mine and plays the role of amplifying the sound by using the smallest bone in the human body. Finally, the inner ear is responsible for balancing sense

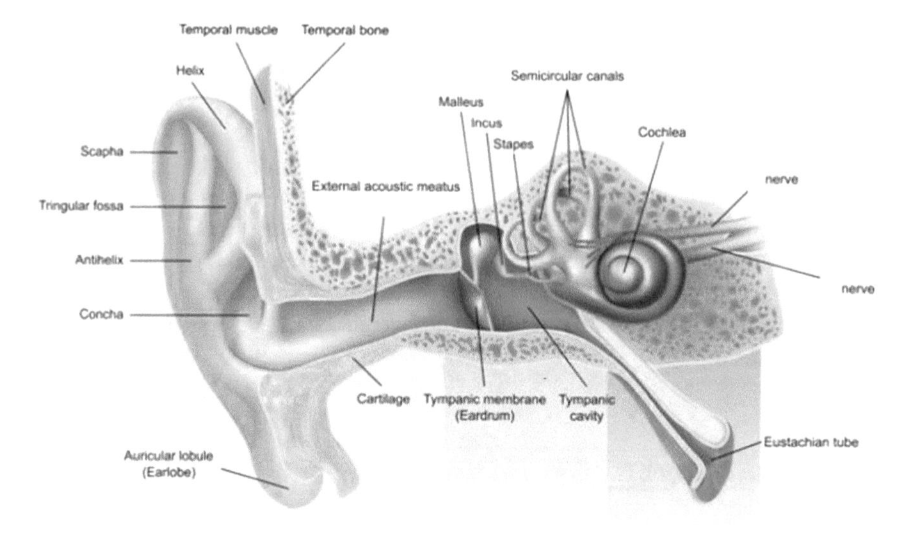

Fig. 1 Ear structure of a person

and auditory functions, and the sound from the inner ear is transmitted to the brain through the auditory nerve [7, 9].

Also, the magnitude of a sound depends on the magnitude of the amplitude, and the larger the amplitude, the smaller the small sound. Figure 2 shows the loudness curve. The sound sensation felt by humans is somewhat different from the physical quantity [9]. The physical quantity is determined by the amount of change in pressure, but the human ear is linearly per decibel (dB). The human ear is capable of recognizing the frequency component of sound. However, it is not perceived to be flat at all frequencies, but is perceived to be a different sensitivity for each frequency band, such as a loudness curve. The size of a sound felt by humans differs from the size of the physical sound because it is a sensory quantity that is perceived by the auditory system [10–12]. The sensible size of the sound is the volume, and the size of the physically measured sound is called the sound pressure, which is the atmospheric pressure change. This is different from the size of the sound felt by human hearing and the size of the physical sound. Therefore, it is necessary to express the magnitude of the sound in units of measurement that are consistent with human hearing. At this time, as the unit of measure used dB (decibel). dB is the logarithm of the physical strength of the two notes, which is the logarithm of the reference value and the measured value [7–13].

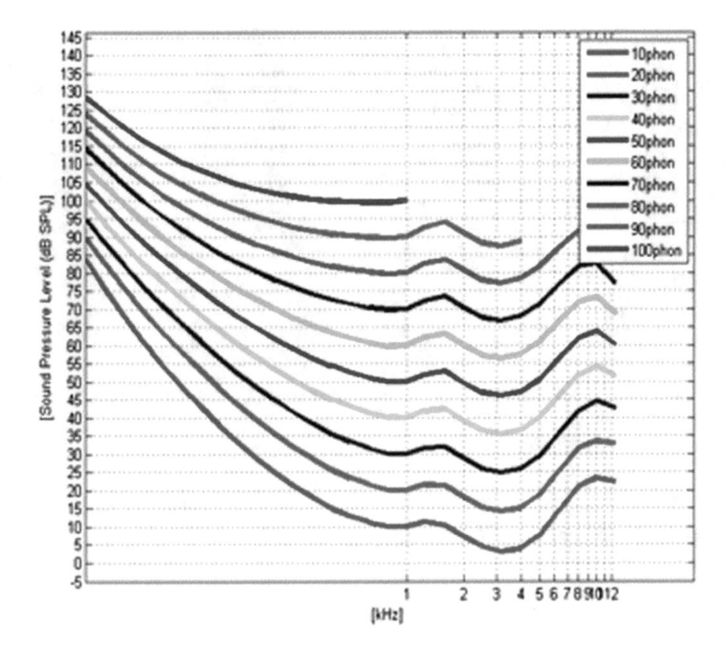

Fig. 2 Typical Equal Loudness Contours [9]

3 Existing Engine Sounds and Virtual Engine Sound

3.1 Original Engine Sound Characteristics

An engine used in an internal combustion automotive engine can be largely divided into a gasoline engine and a diesel engine depending on the method of igniting the fuel. Fuel used in gasoline engines is called gasoline, and fuel used in diesel engines is called diesel. Both fuels are classified according to the process of extraction from crude oil [14]. There are numerous parts that make up the engine of an automobile, but the major parts can be divided into three parts: crankcase, cylinder block and cylinder head. The crankcase has a crankshaft that changes the reciprocating motion of the piston into a rotational motion, and the cylinder block actually forms the skeleton of the engine in which the piston moves. Finally, the cylinder head has an intake valve and an exhaust valve that opens and closes according to the movement of the piston. In the case of a gasoline engine, there is a spark plug. In the case of a diesel engine, there is a fuel injection nozzle. The engine is operated in a four-stroke cycle in the form shown in Fig. 3 below and is divided into an intake stroke, a compression stroke, a combustion stroke, and an exhaust stroke [15].

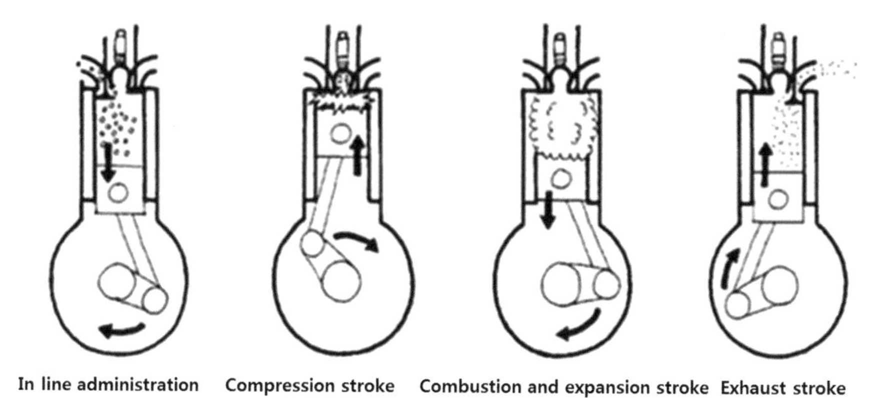

Fig. 3 4 administrative structure

3.2 *Gasoline Engine Characteristics*

Figures 4 and 5 below show the engine sound waveform and spectrum at 2000 RPM of a gasoline engine. The most popular cars on the planet today are gasoline-based cars. A gasoline engine sucks up the fuel mixture and Fig. 5 shows that the first resonance frequency energy is large, at about 65 Hz, and there are several harmonics. (2000 [RPM]/60 [sec]), the resonance frequency that occurs as a pure tone can be seen as the sound generated from the timing belt of the engine, and the cylinder at the highest frequency. After calculating * 2, if the engine is a 4-stroke and division 2 is performed, the result is about 66.6 Hz. It can be seen that the values measured through the sound source are close to the theory [14–16].

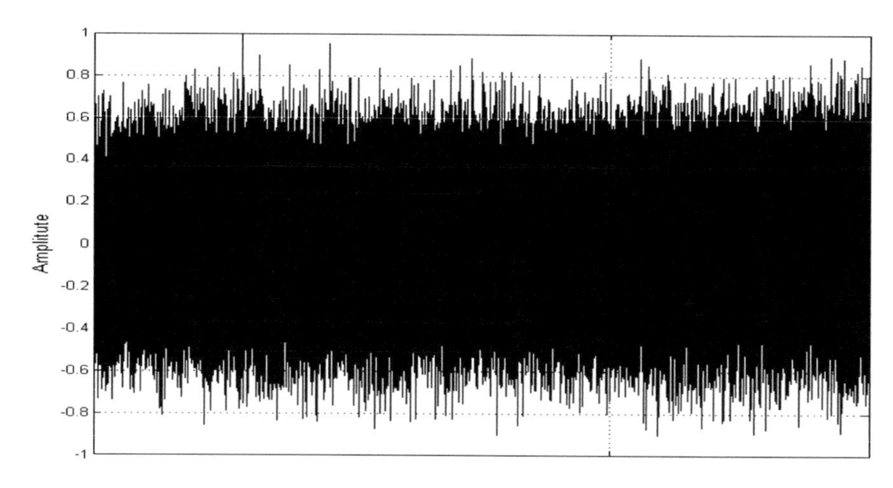

Fig. 4 Gasoline engine sound source waveform at 2000 RPM

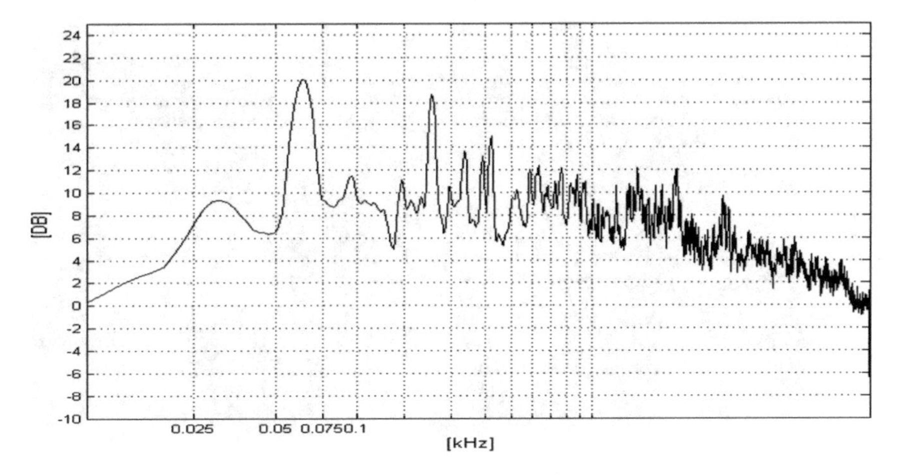

Fig. 5 2000 RPM, gasoline engine sound source logarithmic spectrum

3.3 Virtual Engine Sound

When designing a virtual engine sound, the characteristics of a pedestrian should be considered. Further, there is a need to provide a driving engine sound in consideration of the auditory characteristics for the driver. Figures 6 and 7 show the waveforms and spectrum of the virtual engine sound based on 2000 RPM. Figure 7 shows that the fundamental resonance frequency is about 69 Hz, which is not very different from a conventional engine sound, and the difference in dB is only about 2 dB. However, it was confirmed that the frequency energy in the 1500 Hz band was significantly different and the dB of the harmonic was different. But as a whole, the characteristic of the engine sound followed. The reason why a virtual engine sound and the characteristics of a conventional engine sound should be similar is because when a person is a familiar with an engine sound heard since birth, a virtual engine sound can be mistaken as a sound other than an engine sound [14–17].

4 Experiments and Results

In this paper, to study the acceleration of a virtual engine sound according to the change of the playback speed based on a specific speed (4000 RPM) standard, the playback speed of a virtual engine sound was set to 1, 2, 4 and 5 s. The characteristics of each of was analyzed, and the acceleration-MOS test was based on these characteristics.

The experiment was prepared by quantizing the synthesized virtual engine sound based on the engine sound of a gasoline vehicle (Kia Motors 2010 FORTE model) at 16 bit/sample. Also, in order to confirm the section where acceleration was high,

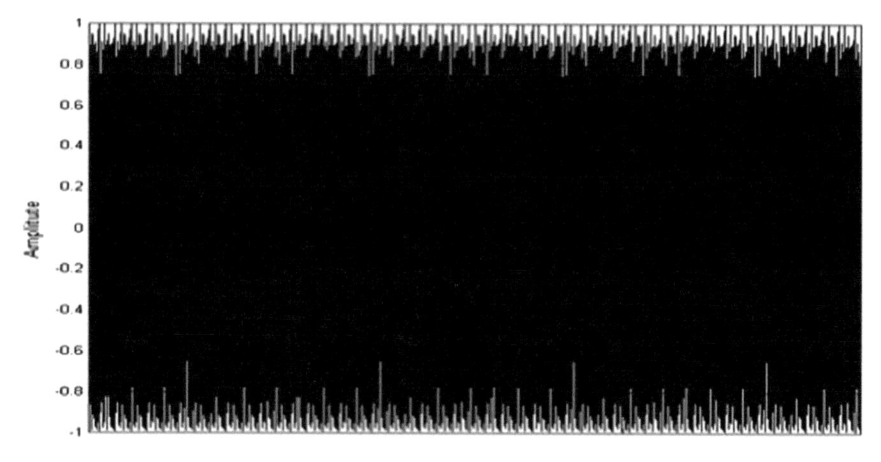

Fig. 6 Virtual engine sound source waveform at 2000 RPM

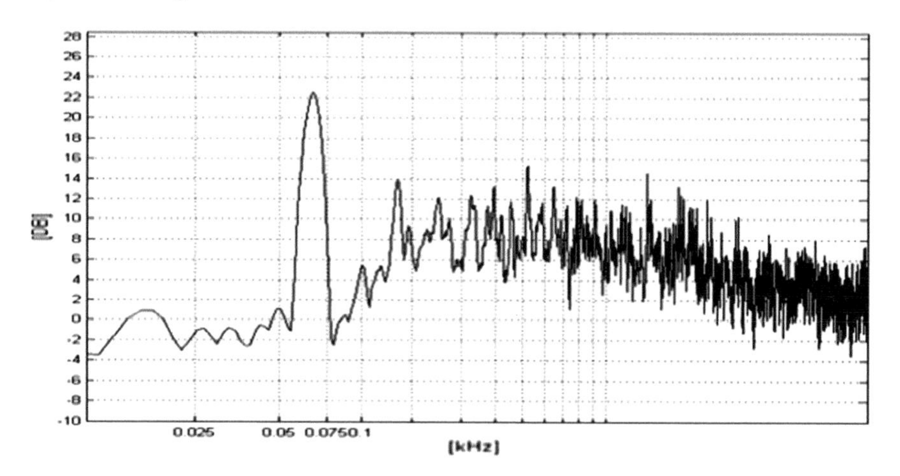

Fig. 7 2000 RPM, virtual engine sound source logarithmic spectrum

the experiment was performed at 4000 RPM, which increased when accelerating. Also, a comparative analysis in the frequency domain and the spectrogram domain of these digital materials were performed. In addition, the MOS test was carried out with the sound source converted from the playback speed, and the most accelerated virtual engine sound was selected. Figures 8 and 9 show the 4000 RPM waveform and logarithmic spectrum of the virtual engine sound. Figure 9 shows that the resonance frequency energy was large, at about 102 Hz, and the harmonic was remarkable as compared at 2000 RPM.

Figure 10 is a spectrogram showing the reproduction time of the virtual engine sound source. Each sound source was quantized at 16 bit/sample at 13000 Hz. In order to test the acceleration, 3 s were converted into 1, 2, 4, and 5 s. When compressed for acceleration, energy was distributed evenly from 400 Hz to 4600 Hz in 1 s. However,

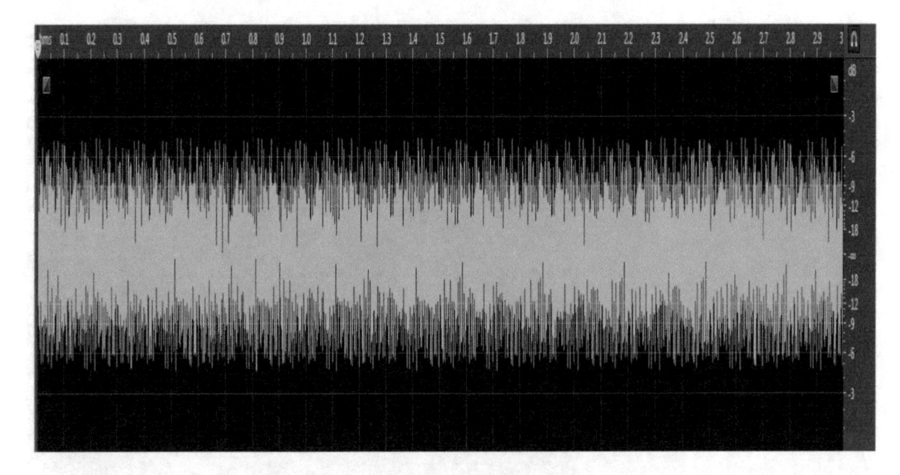

Fig. 8 Virtual engine sound source waveform at 4000 RPM

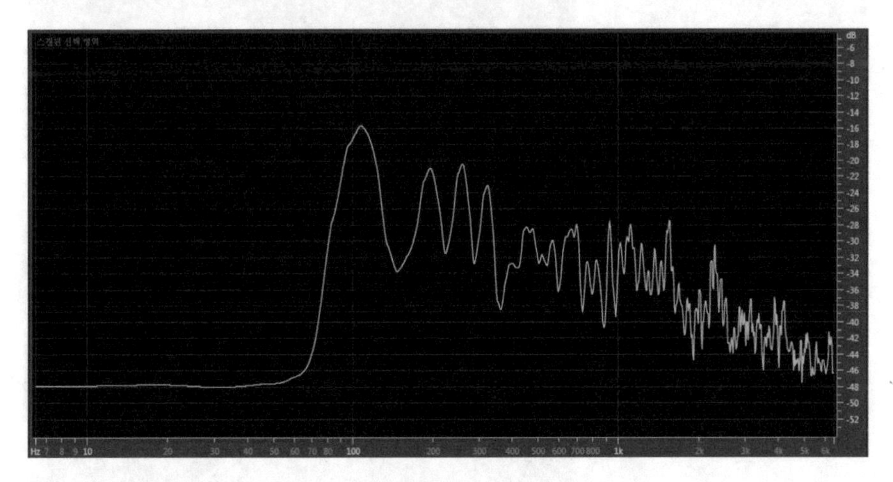

Fig. 9 4000 RPM, Virtual engine sound source logarithmic spectrum

as the time increased, the distribution of the energy converged to a low frequency, and the high frequency energy component decreased. Also, when the time exceeded 4 s, almost no high frequency energy was found, and energy was concentrated from 100 Hz to 2600 Hz. In the case of 5 s, the energy was distributed only up to the 2000 Hz band, but the energy density was higher than other times. If you look at these features, adjusting the playback speed from 1 to 3 s can cause people to feel that they are not well because they are too sensitive to sounds. Energy is concentrated in the sensitive zone of the human ear for 4–5 s, so even if the high frequency energy was small, it could sufficiently be recognized as an engine sound.

Figure 11 shows the frequency comparison analysis for each hour: 1 s for red, 2 s for orange, 3 s for yellow, 4 s for green, and 5 s for blue. The faster the playback

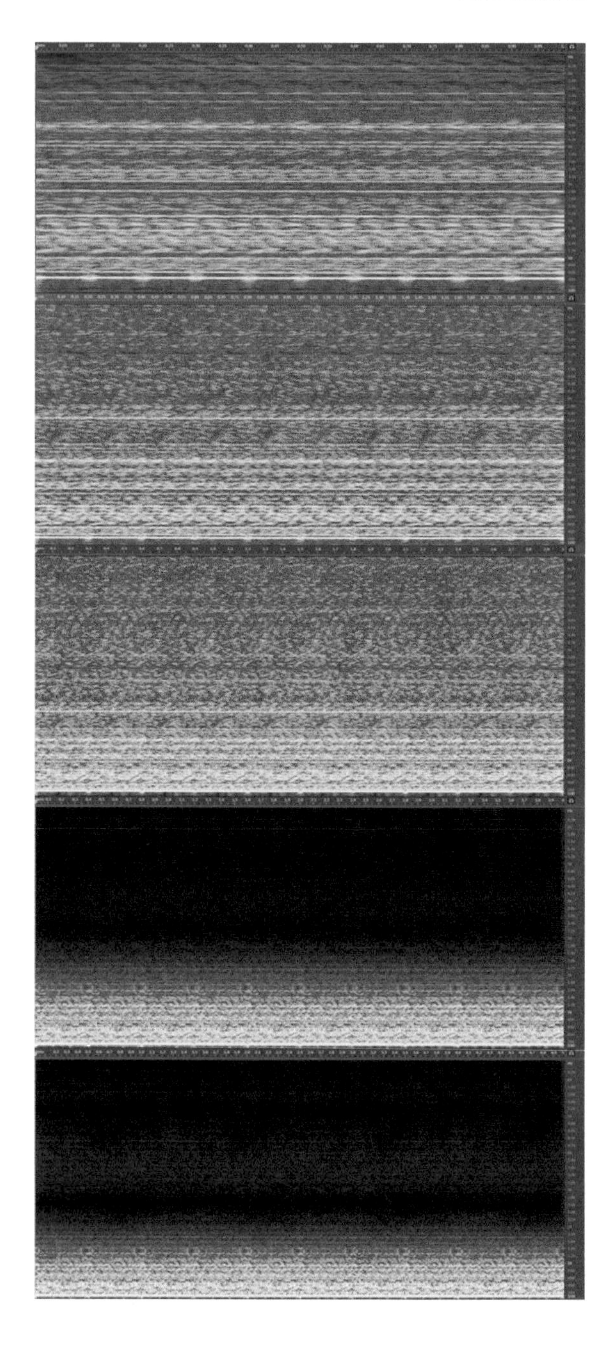

Fig. 10 A spectrogram analysis that converts the sound of an existing 4000RPM engine sound into 1, 2, 3, 4, and 5 seconds, respectively, through time domain transformation

speed, the higher the fundamental resonance frequency, and the longer the playback time, the lower the resonance frequency. Also, at 4 and 5 s, the energy of the higher frequency suddenly dropped, and the falling slope from 1 to 3 s was very different.

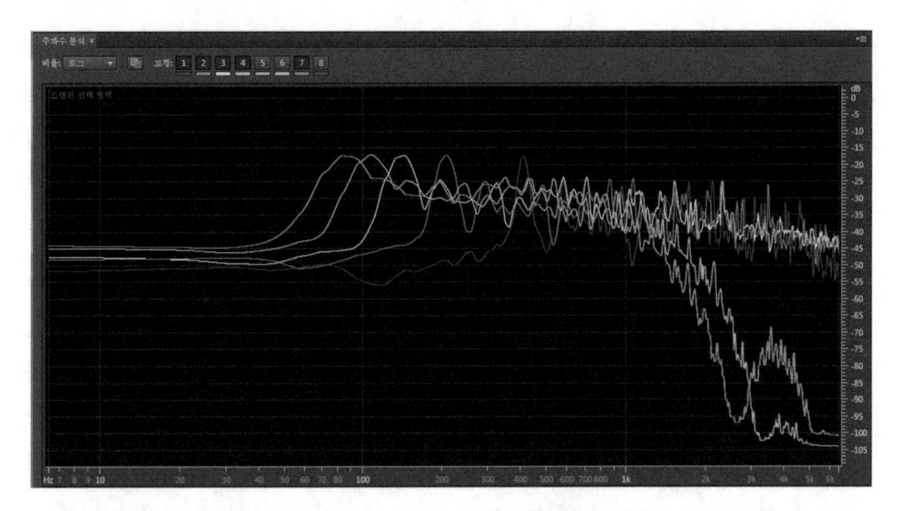

Fig. 11 4000 RPM, Virtual engine sound source logarithmic spectrum

Figures 12 and 13 show the results of the experiment with 10 males between 20 and 30 years old who directly drove to perform the acceleration test of each sound source. A MOS test was conducted. Figure 12 shows that an engine sound of 4 s can make subjects feel virtual acceleration. In Fig. 13, 90% of the respondents answered that they could recognize the engine sound, while 10% did not.

5 Conclusion

In the age of 5G, the development of eco-friendly cars is very important. In particular, eco-friendly automobiles under active research and development are expected to lead the next generation automobile market together with autonomous vehicles. However, in the case of an eco-friendly car, an artificial engine sound (VESS) is required because of quieter actual engine sounds. In developed countries, it is obligatory to insert a virtual engine sound into an environmentally-friendly automobile. However,

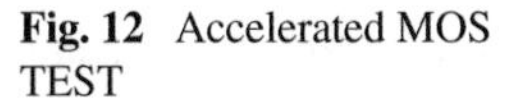

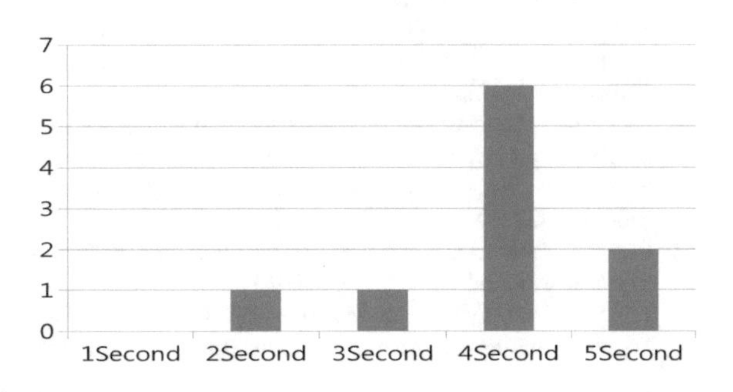

Fig. 12 Accelerated MOS TEST

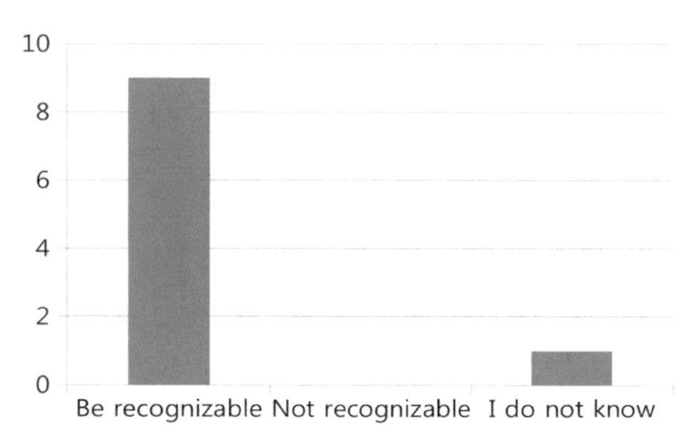

Fig. 13 MOS TEST of whether virtual engine sound is recognized as engine sound

not only the engine sounds are inserted, but also there is a part of the driver who cannot feel the engine sound felt by the Pre-learning and listening effect for the time being. In this study, we studied the change of the acceleration engine sound output when the driving speed of an eco-friendly automobile was increased by varying the time. Experimental results show that the compression and output of an existing virtual engine sound (3–4 s) are the most accelerated. In the future, it is hoped that each virtual engine sound, according to the type of car and exhaust, will help prevent damage to property and humans caused by traffic accidents.

References

1. Myung, D.K., Nam, S.W., Pho, J.H. (eds.): An Active Noise Control of Actual Diesel Engine Exhaust System. The Korean Society for Noise and Vibration Engineering (2013)
2. Kang, B.M., Son, Y.D. (eds.): Design of Virtual Engine Sound System for Green Car. The Korean Institute of Electrical Engineers, pp. 93–94 (2015)
3. Bae, M.J., Lee, S.H. (ed.): Digital Voice Analysis. Korea
4. Jee, S.H., Park, H.W., Bae, M.J. (eds.): A Study on Virtual Engine Sound Design. The Acoustical Society of Korea, Vol. 35, p. 107 (2016)
5. The Korean intellecual Property Office, VESS (Vurtual Engine Sound System). Num, 10-2014-008317
6. Park, H.W., Jee, S.H., Bae, M.J. (eds.): Virtual-engin sound design through internal combustion engine acoustics analysis. Asia-Pacific J. Multimedia Serv. Convergent with Art, Humanit. Sociol **6**(11), 649–656 (2016)
7. Bae, M.J., Lee, S.H. (eds.): Digital Voice Analysis. Korea
8. Lawrence, R.R., Schafer, R.W. (ed.): Theory and applications of dig digital speech processing. PEARSON (2011)
9. ISO 3745. Acoustics—determination of sound power levels of noise sources using sound pressure—precision methods for anechoic and hemi-anechoic rooms (2003)
10. Zwicker, E., Fastl, H.: Psychoacoustics—facts and models. Springer (1990)
11. Moore, B.C.J.: An introduction to the psychology of hearing, 5th edn.
12. Cha, J.S., Lee, S.D., Kang, S.G., Lee, K.W., Chen, K.W., Jung, K.S., Park, S.K.: J. Korean Soc. Mar. Environ. Saf. **19**, 658–665 (2013)
13. Jee, S.H., Park, H.W., Bae, M.J. (eds.): A Study on Imaginary Engine Sound with Same Sound Width by Frequency Direction. The Institute of Electronics and Information Engineers. (2016)

14. Jee, S.H., Park, H.W., Bae, M.J. (eds.): A study on virtual car engine sound design according to sine wave harmonics frequency bandwidth. The Korean Inst. Commun. Inform. Sci. **60**, 1527–1528 (2016)
15. Park, H.W., Jee, S.H., Bae, M.J. (eds.): Virtual car engine sound classification study of the band SB. Convergence Res. Lett. **2**(3), 787–790 (2016)
16. Bae, S.G., Lee, W.H., Bae, M.J.: A study on low frequency noise of dehumidifier using acoustic characteristics. Int. J. Eng. Technol **8**, 235–237 (2016)
17. Park, H.W., Lee, W.H., Bae, M.J.: Annoyance study about exposure time due to megaphone noise. In: 2nd International Conferrence, ICCPND 2015, vol. 5, pp. 19–20 (2015)

A Study on the Characteristics of an EEG Based on a Singing Bowl's Sound Frequency

Ik-Soo Ahn, Bong-Young Kim, Kwang-Bock You and Myung-Jin Bae

Abstract We analyzed the sound of a singing bowl, which is used as a method to restore and maintain the balance of the natural frequency of the human body, and studied the EEG (electroencephalogram) of the listener according to the frequency band of the singing bowl's sounds. The singing bowl's sound has a wide range of frequencies that can restore all of the natural frequencies of the human body, thus helping the body recover or heal. The low frequency band of the singing bowl's sound was the most energy-intensive across the whole band, followed by a decrease in energy in the order of the mid-frequency band and the high-frequency band. The singing bowl is a tool designed to help people heal and recover, thus it's sounds aim to gives stability and comfort to people's mind and body as a whole. A study on the brain wave characteristics of listeners corresponding on the frequency bands of a singing bowl's sounds will provide useful data for using singing bowls and it is hoped that this study will contribute to the development of other sound healing tools.

Keywords Sound · Human body · Influence · Singing bowl's sound · Healing Frequency · EEG

I.-S. Ahn · B.-Y. Kim · K.-B. You
Department of Information and Telecommunication, Soongsil University, Dongjak, Seoul, South Korea
e-mail: aisgoodman@ssu.ac.kr

B.-Y. Kim
e-mail: bykim8@ssu.ac.kr

K.-B. You
e-mail: kwangbockyou@ssu.ac.kr

M.-J. Bae (✉)
Department of Information and Telecommunication Engineering, Soongsil University, Dongjak, Seoul, South Korea
e-mail: mjbae@ssu.ac.kr

R. Lee (ed.), *Software Engineering Research, Management and Applications*, Studies in Computational Intelligence 789, https://doi.org/10.1007/978-3-319-98881-8_16

1 Introduction

In modern quantum physics, the earth and the natural frequencies of the human body are said to be maintained by balancing the sounds of nature, the sounds of various tools, and music. From this point of view, studies on the effects of sound on the human body have been demonstrated by various methods. Practically, there are many cases in which music or sound affects the human body and brings healing or therapeutic effects. Among them, singing bowls, which first appeared in Nepal, Tibet and India, are mysterious bowl-shaped tools that produce sounds and have many healing and therapeutic effects on the human mind and body. In this paper, to reduce the scope of potential controversy with the medical field, we will use the term "healing" rather than the term "treatment" as a unified explanation. A singing bowl is a bowl-shaped tool made primarily of bronze and tin which produces a sound similar to that of Korea's Bangjja-Yugi. A singing bowl consists of several bowls, which generate various frequencies depending on size and thickness. The sound generated by a singing bowl creates a natural frequency which works on the brain to stimulate emotions, comfort, serenity, or to induce brain activity. In this paper, we investigate how the singing bowl's sounds emotionally influence an EEG. This study on the EEG characteristics of the singing bowl's sounds will play a role in confirming the healing range and effectiveness of various sound healing tools as well as the singing bowl. To investigate the EEG characteristics of the singing bowl's sounds, we first extracted the singing bowl's sounds and their overall mean frequency. Next, a singing bowl's sounds were presented to listeners so that their brain waves could be measured. The waves of the listeners hearing the singing bowl's sounds were divided into the frequency bands of the singing bowl's sounds. The results of this study will be based on emotional response data in the use of a singing bowl's sounds and it will form the outlines of a manual to use singing bowls more effectively. We also hope that it can be used as valuable data for future research and verification of useful sound healing tools [1–3].

2 Principle of Occurrence of the Singing Bowl's Sound

The singing bowl is an alloy body made of a mixture of seven metals (gold, silver, iron, mercury, tin, copper, lead), and the main component is a bowl-shaped object made of brass and tin used for strength. Singing bowls can also be used as bowls, but have been used primarily as a tool for sound healing. To play a singing bowl, a singing bowl stick is rubbed or hit against the singing bowl to create a vibration frequency. The sound made by rubbing is soft and gentle, so it resonates continuously. The sound generated by tapping is stronger, and it repeats until it gives a lull and disappears. Thus, a singing bowl's sound can be strong or weak, rough or soft depending on which of two methods is used. The methods of producing a singing bowl's sounds need to be continuously developed in accordance with the healing site and the situation of

(a) Singing bowl and Stick

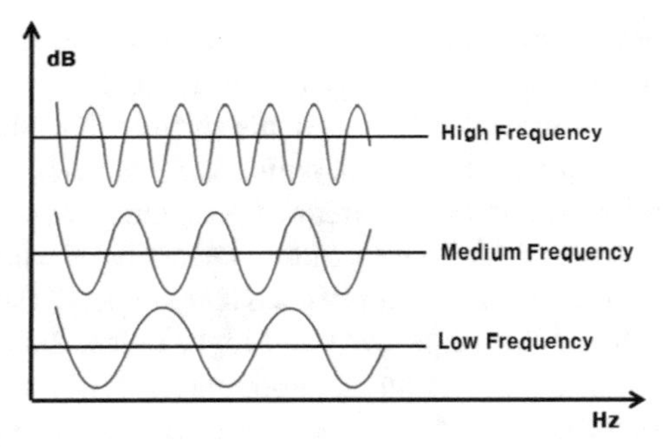

(b) Possible frequency of singing bowl's sound

Fig. 1 Sound range of singing bowl's sound

the person being healed, and used in such a way as to accurately generate the various required frequencies every time [4–6] (Fig. 1).

The frequency of the singing bowl's sounds range from low to high frequencies, including sounds not perceptible to human hearing. However it produces sounds ranging from very low frequencies to very high frequencies that the body perceives. The number and placement of singing bowls to be used depends on the type of healing to be done. A single singing bowl can be consistently used to heal one part of the body. However, when healing the whole body, including the mind, a dozen of singing bowls as one set are used depending on the exact context. The singing bowls are placed around the body of the person to be healed from the head to the base of the feet. The person to be healed sits or lies down in a relaxed posture, and the singing bowls are placed on both sides of the person's body. Sometimes a singing bowl is placed on each part of the body. In order to heal the internal organs of the human body, a relatively large singing bowl is used to produce long-lasting bass. In order to heal the brain, skeleton, and skin, singing bowls of the size appropriate for each natural frequency are used [7–9].

3 Frequency Analysis of Singing Bowl's Sound

In order to analyze the frequency of a singing bowl's sounds, a singing bowl used for comprehensive healing of a range of conditions was selected, and the sounds it made were recorded and analyzed with an acoustic analysis tool. The environment for the acoustic analysis tool was analyzed by spectral graph with the sample rate set to 3200 Hz, the channel set to mono, and the resolution set to 16 bits. The frequency characteristics of the Singing bowl's sounds were divided into three parts. The first part was the low frequency region below 1000 Hz, the second part was the mid-frequency region from 1000 to 4000 Hz, and the third region was the high-frequency region above 4000 Hz. As shown in the frequency spectrum graph of the singing bowl's sounds in Fig. 2 below, the first region, the low frequency region, forms the widest bandwidth and generates a sound that can restore the fundamental frequency of the human body in a stable manner. The second mid-frequency region forms a number of distinct peak points and generates a frequency that energizes the human brain. The third high-frequency region shows a calm pitch graph and gives comfort to human brain waves, further enhancing the healing effect [10, 11].

The frequency graph of the singing bowl's sounds shows that the frequency band below 1000 Hz is wide and high. Below 1000 Hz there is a low-frequency band that can tune the human body into it's vibrations. In particular, it restores natural frequencies to the skin and internal organs of the body and in so doing heals the body. Next, several distinct waveforms are seen in the region from 1000 to 4000 Hz. These waveforms are psychologically concentrated and refreshing, making them more effective for emotional and mental healing. In addition, the singing bowl's sounds includes both ultra-low frequency and very high frequency ranges, which affect overall human skin and the internal organs and stomach. In this way a singing bowl's sounds affect both the mind and body of a person and provides healing and therapeutic benefits [12–14].

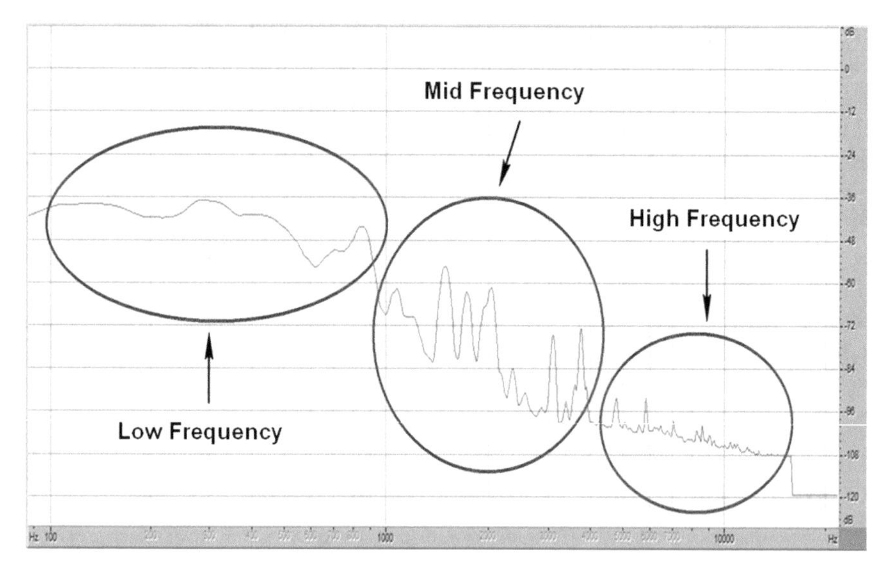

Fig. 2 Frequency spectrum graph of singing bowl's sound

4 EEG Characteristics by Frequency of Singing Bowl's Sound

Singing bowl's sound is a mysterious and elaborate sound that causes the healing and healing effects of a person's mind and body. To investigate human emotional changes in the frequency range of the singing bowl's sound, we analyzed the human brain waves responding to each frequency range of the singing bowl's sound.

4.1 EEG Characteristics

An EEG is a method to record electrical activity in the brain. It works by continuously recording potential electrical fluctuations between two points on the scalp caused by electrical activity of the brain. In 1875, the British physiologist R. Cayton first reported on the system of weak electrical activity from the cerebral cortex of rabbits and monkeys. The first measurement of human brain waves was by German psychiatrist H. Berger in 1924. Brain waves flow from every person's brains at every moment of their existence. An EEG refers to the flow of electricity that occurs when a signal is transmitted between brain nerves. Brain waves are classified into delta waves, theta waves, alpha waves, beta waves, and gamma waves depending on the activity state of the brain. The delta (δ) wave is divided into a vibration frequency of 0.2–4 Hz, a theta (θ) wave of 4–8 Hz, an alpha (α) wave of 8–13 Hz, a beta (β) wave of 13–30 Hz and a gamma (g) wave of 30–50 Hz. Delta waves occur when you are in a state of deep sleep, meditation, unconsciousness, coma or anesthesia. The theta wave appears mainly in emotionally stable states or in drowsiness that enters into sleep and is more common in children than in adults. Alpha waves appear when people are awake and maintaining a calm mental state. The more stable and relaxed they are, the more the amplitude increases. In general, it appears continuously in the form of regular waves, with the largest recorded at the top and back of the head and the smallest at the front of the head. Stable alpha waves will appear when a person has their eyes closed, and is in a state of relaxation. However they will be suppressed if that person opens their eyes and observes an object or becomes mentally excited or agitated. Beta waves appear predominantly at the front of the head, appearing when performing all conscious activities, such as when a person is awake or speaking. High beta waves may be strong when handling anxiety, tension, excitement, stress, and complex calculations. Gamma waves appear to oscillate more rapidly than beta waves, appearing when a person is emotionally very irritated, or excited, unstable, or using reasoning, logic or judgment [15, 16] (Table 1).

Table 1 Types of EEG

EEG	Frequency (Hz)	Characteristic
Delta(δ)	0.2–4	Sleep, meditation, unconsciousness or coma, anesthesia
Theta(θ)	4–8	Emotional stable state, drowsy state or shallow sleep state
Alpha(α)	8–13	An arousal state that maintains a mental state, Stable and relaxed state, closed eyes and true state
Beta(β)	13–30	When awake, when speaking, when doing all the conscious activities High beta waves (18–30 Hz) are strong when handling anxiety, tension, excitement, stress, and complex calculations
Gamma(g)	30–50	Emotionally very irritable, excited, uneasy, When you are devoted to logic learning such as reasoning and judgment

4.2 Analysis of EEG Characteristics by Frequency Band of Singing Bowl's Sound

A singing bowl makes a mysterious and subtle sound that stimulates the mind and body of a person and helps them heal and gives them therapeutic benefits. A singing bowl's sounds are conveyed along the human audible frequency band (20 Hz–20 kHz) as well as ultra-low frequency and ultra-high frequency sounds which the human body perceives through the skin or internal organ framework. In order to investigate the response of people to the various frequencies of sounds emitted by a singing bowl, an EEG was measured and analyzed. Analysis of the EEG characteristics of the singing bowl's sound frequency band will have a great influence on the wide use and popularization of singing bowls, and will be useful reference material for refining and developing the usage method of singing bowls. In order to investigate human responses to the various frequencies of singing bowl's sounds, we divided the frequency into low frequency, mid-frequency, and high-frequency bands. In order to measure the EEG response to the singing bowl's sounds, three men and 20 women in their twenties were asked to listen to a singing bowl's sounds. The average EEG distribution was sampled and collected as the data used in this paper. After listening to the singing bowl's sounds, the waveforms showed different frequencies according to the frequency of the singing bowl's sounds. Gamma, Delta, and Theta waves were greatly increased in the participant's brain waves who listened to the low-frequency sounds from the singing bowl. In particular, gamma, delta, and theta waves of the right brain reflecting emotional changes showed a sharp and rapid increase. This phenomenon shows that the singing bowl's sound helps the participants to heal and gave them therapeutic value by creating emotional stability and mental well-being through these gamma, delta, and theta waves. Gamma, delta, and theta waves of the participants listening to the mid-frequency bands of the singing bowl were slightly smaller than those actually emitted by the singing bowl, although the rise was smaller than that of the low frequency bands (Fig. 3).

This phenomenon helps maintain a comfortable and stable EEG characteristic of gamma, delta, and theta waves that assist in healing and have therapeutic value even in the mid-frequency band. In the left brain, the rise of gamma and delta waves was similar to that of the mid-frequency band, as reflected in the energy of the high frequency band of the singing bowl's sounds. However, in the right brain the EEG responded more to emotion. Alpha waves showed a significant increase, the gamma and beta waves were greatly reduced, and the delta waves and theta waves were significantly reduced. This phenomenon is attributed to the fact that a singing bowl's sounds are focused on the low-frequency range and is a tool for healing through meditation and helping parts of the body such as skin and muscles [17–20].

- Before listening to the singing bowl's sounds, brain waves showed no significant change from the standard frequencies of gamma, delta, theta, alpha, and beta waves (Fig. 4).
- After listening to the low frequency sounds of the singing bowl, brain waves showed a rise in gamma and delta waves, and the sensation-related right brain showed a large increase and a steady state of drowsiness and sleep state (Fig. 5).
- After listening to the singing bowl, brain waves showed a slightly relaxed and stable state with a slight increase in gamma, delta, theta, alpha and beta waves (Fig. 6).
- After listening to the singing bowl, brain waves showed a steady and relaxed awakening, particularly in the right hemisphere of the brain that controls emotions.

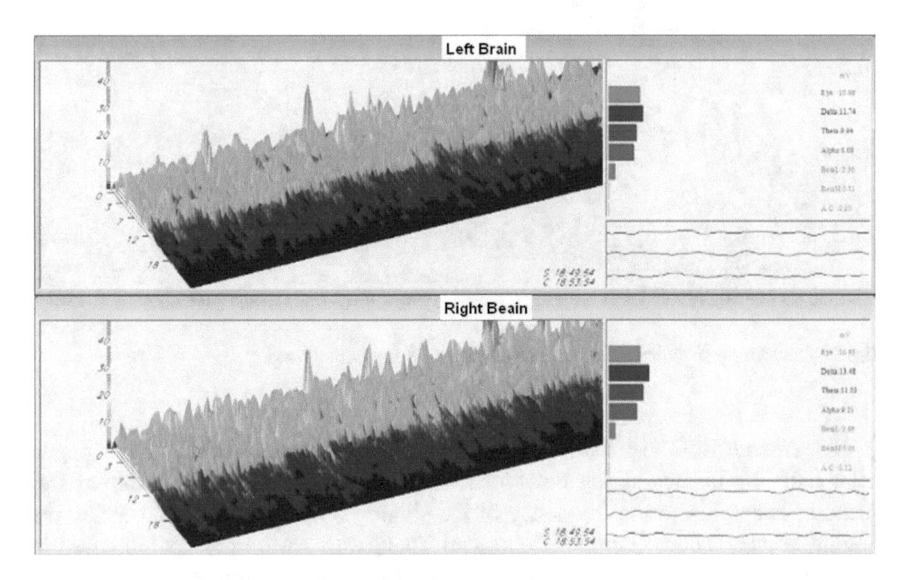

Fig. 3 The EGG distribution before listening to the singing bowl's sound

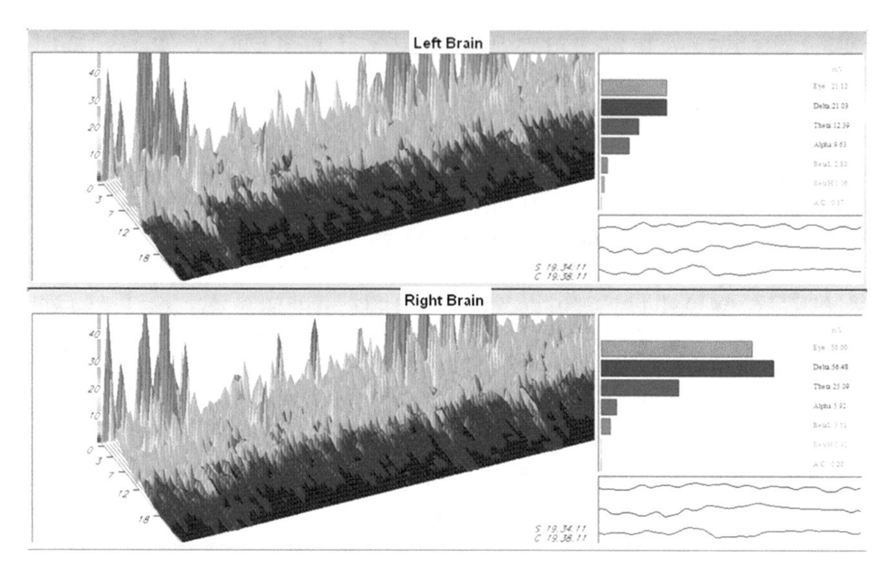

Fig. 4 The EGG distribution after listening to the low frequency band

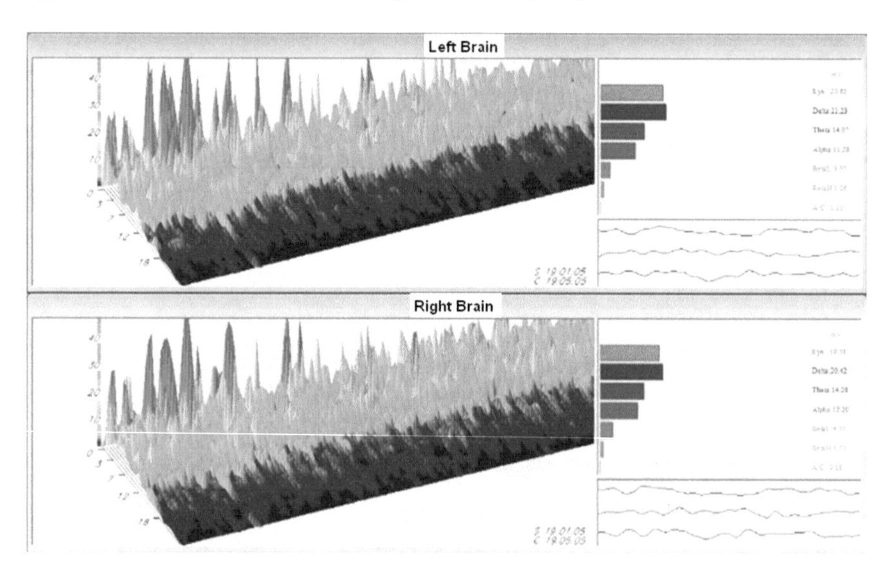

Fig. 5 The EGG distribution after listening to the mid frequency band

The average EEG distribution was sampled and collected as data of this paper. After listening to the singing bowl's sound, the waveforms showed different frequencies according to the frequency of the singing bowl's sound after hearing the singing bowl's sound. Gamma, Delta, and Theta waves were greatly increased in the celadon's brain waves that listened to the low-frequency sounds of the singing

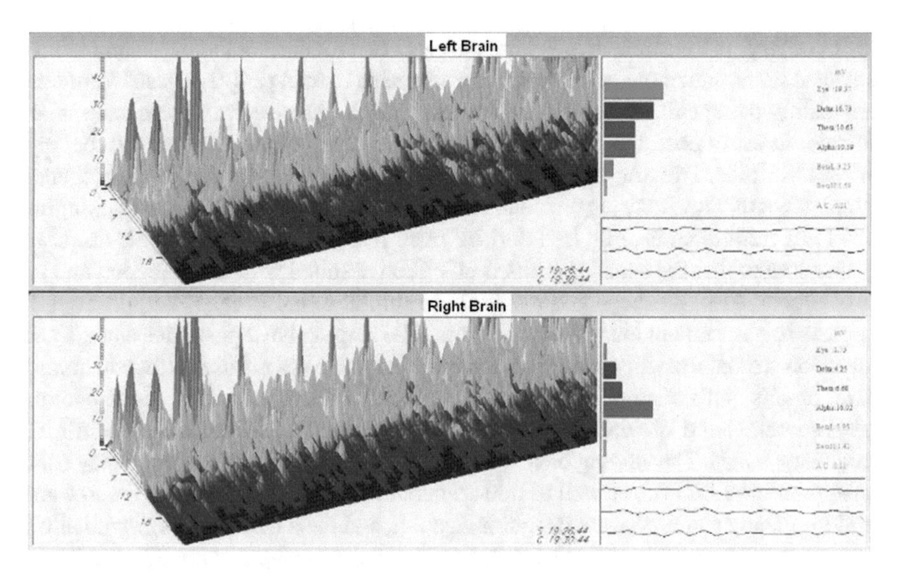

Fig. 6 The EGG distribution after listening to the high frequency band

bowl's sound. Especially, gamma, delta, and theta waves of the right brain showing emotional changes showed a sharp and rapid increase. This phenomenon shows that the singing bowl's sound helps the celadon to heal and therapeutic by creating emotional stability and mental well-being of gamma, delta, and theta. Gamma, delta, and theta waves of the celadon listening to the mid-frequency bands of the singing bowl's sound were slightly smaller than those of the singing bowl's sound, although the rise was smaller than that of the low frequency bands. This phenomenon maintains the comfortable and stable EEG characteristics of gamma, delta, and theta waves that help in healing and therapeutic even in the mid-frequency band. In the left brain, the rise of the gamma and delta waves was similar to that of the mid-frequency band, as reflected in the energy of the high frequency band of the singing bowl's sound. However, in the right brain EEG responding more to emotion, Alpha wave showed a significant increase, the gamma and beta waves were greatly reduced, and the delta waves and theta waves were significantly reduced. This phenomenon is attributed to the fact that singing bowl's sound is focused on the low-frequency range and is a tool for healing through meditation and helping the body such as skin and muscles [9, 10].

5 Conclusion

The often harsh environment of modern civilization can destroy the human rhythm and cause stress and diseases. However, such stress or diseases can still not be cured completely by modern medicine. Modern medicine focuses more on therapy, and uses surgery, medication or physical contact. In Oriental medicine, acupuncture, physical therapy, and medicinal herbs are also used to treat people. These treatments are mainly therapeutic concepts, and to date have not received extensive national or social interest in post-treatment sequelae, mental and physical healing and therapy, at least in Korea. The singing bowl covered in this study is a new health maintenance supplement therapy focusing on mental and physical healing and therapy. The singing bowl is a traditional healing tool derived from India, Tibet, and Nepal. It has been demonstrated that the sound it emits is an effective sound with low rejection and no side effects. Also, its shape and material are similar to the Korean Bangjja-Yugi. It appears to have no obvious negative points. The singing bowl generates sounds and vibrations to stabilize the emotional state of mind and has a positive effect on human skin, organs and skeleton. In this paper, we studied the frequencies of the singing bowl's sounds and divided them into low frequency, middle frequency, and high frequency bands. The singing bowl's sound strongly generates low - frequency band energy below 1,000 Hz, as well as mid-frequency energy. The energy of the low and mid frequency bands activate the gamma, delta, and theta brain waves, which allow healing and therapy for the human mind and body by making emotionally stable and comfortable conditions suitable for meditation. This study will hopefully lead to the development and validation of various healing and therapeutic tools in the future, and also shows that we need to further open up the field of human health to viable alternative medicines, as many of the ailments we face cannot be solved by modern medicine alone.

References

1. Ryu, J.-S.: A comparative study on the effectiveness of surface electromyography in the treatment of shoulder cuffs using magnets, tuned ankles and clans. Master's Thesis, Kyonggi University (2008)
2. Lee, C.-S.: Eastern and Western Therapy for Sound (monthly) Practical Special Education (2015)
3. Shi-a Chun Station, Shuren Shrestha: Singing Bowl Healing. Genbook (2015)
4. Jeon, H.-S.: J. Korean Soc. Hortic. Sci. Technol. (2008)
5. Choi, H.-S.: The effect of sound therapy on training therapy. Master Thesis, Yonsei University (2007)
6. Kim, Y.-W.: Comparison of THI Changes According to the Duration of TRT, Sound Therapy Tools, and Lingual Counseling. Hallym University Graduate University (2009)
7. Lee, Y.-H.: What is Sound Therapy? Academic papers, Korea Occupational Health Association (2003)
8. Bae, M.-J., Lee, S.-H.: Digital Speech Analysis (1998)
9. Rabiner, L.R., Schafer, R.W.: Theory and Applications of Digital Speech Processing. Pearson (2010)
10. Kang, S.H., Jung, S.H., Jung, H.K., Lee, J.W. (eds.): Analysis of Reverberation time. The Acoust. Soc. Korea, 311–314 (2010)
11. Yun, B.H., Baek, G.R., Bae, M.J. (eds.): A study on building SVDB to monitor a soft voice. Inst. Electron. Inf. Eng. 800–801 (2011)
12. Park, H.W., Bae, S.G., Bae, M.J. (eds.): Improving pitch search through emphasized harmonics. The Acoust. Soc. Korea, 230–232 (2012)

13. Ahn, I.-S., Bae, M.-J., Bae, S.-G.: A study on promoting appetite in sound signal processing. MAGNT Research Report **2**(5), 105–109 (2014)
14. Lee, W.-H., Bae, S.-K., Bae, M.-J.: A study on low frequency noise analysis of dehumidifier using acoustic characteristics. In: Proceedings of 31st Annual Conference on Voice Communication and Signal Processing, vol. 145–146 (2014)
15. Hall, H., Jeon, J.-Y.: Allies, Do you cure illness with sound? The Sea Publishers, (Korea) Skeptic: promoting science and critical thinking vol. 1, pp. 8–15(2015)
16. Brannon, L., Paste, J., Han, D.-W.: Health Psychology. Sengei Learning Korea (2011)
17. Batsell W.R.,·Brown A.S.: Human flavor-aversion learning: a comparison of traditional aversions and cognitive aversions. Learn. Motiv. **29**:383–396 (1998)
18. Bae, Seong-Geon, Bae, Myung-Jin: A study on recovery in voice analysis through vocal changes before and after specch using speech signal processing. IJAER **12**, 5299–5303 (2017)
19. Bae, S.G., Kim, M.S., Bae, M.J.: On enhancement signal using non-uniform sampling in clipped signals for LTE smart phones. In: 2013, IEEE ICCE-Berlin, pp. 125–126. ICCE-Berlin (2013)
20. Bae, M., Kim, M.: Professor Bae's Sound Story. Gimm-Young Publishers, Inc (2013)

A Study on the Stability of Ultra-High Frequency Vocalization of Soprano Singers

Uk-Jin Song, Ik-Soo Ahn, Myung-Sook Kim and Myung-Jin Bae

Abstract Among the myriad of musical activities, singing, that is, playing with the human voice, has a longer history than any other instrument, and it still draws the attention of a large portion of the audience who listens to music today. Singers cover a wide range of fields and areas of expertise, including popular music singers, Korean traditional musicians, and classical vocalists. In general, when we sing, our voice clearly shows vibrations at low and mid-range frequencies, but these vibration characteristics deteriorate in the high-frequency range. There are, however, a few female sopranos who can consistently produce the highest notes even in the high frequency range. In this paper, in order to confirm whether these soprano singers actually show distinct vibrations in the high frequency range, we analyze the vibration characteristics of four Korean sopranos to ascertain the depth of their vocal vibration. The results show that the soprano singer D has the highest vibration depth, followed by soprano A, soprano C, and soprano B.

Keywords Song · Vibration · Fundamental frequency · Vibration depth
Vocalist soprano

U.-J. Song
Sori Engineering Lab, Soongsil University, DongJak-Gu, Seoul, South Korea
e-mail: imduj@ssu.ac.kr

I.-S. Ahn
Sori Engineering lab, Department of Information and Telecommunication Engineering, Soongsil University, Seoul, South Korea
e-mail: aisbestman@ssu.ac.kr

M.-S. Kim
Department English Language and Literature, SoongSil University, Sangdo-Dong, DongJak-Gu, Seoul, Korea
e-mail: kimm@ssu.ac.kr

M.-J. Bae (✉)
Sori Engineering Lab, Department of Telecommunication Engineering, Soongsil University, Sangdo-Dong, DongJak-Gu, Seoul, South Korea
e-mail: mjbae@ssu.ac.kr

R. Lee (ed.), *Software Engineering Research, Management and Applications*, Studies in Computational Intelligence 789, https://doi.org/10.1007/978-3-319-98881-8_17

1 Introduction

Has a longer history than playing any other instrument, and it still draws the attention of a large portion of the audience who listens to music today. Singers cover a wide range of fields and areas of expertise, including popular music singers, Korean traditional musicians, and classical vocalists. One of the representative elements of singing is vibration. Vibrations impart longevity and maturity of sound to the listener, and it has been confirmed by much research that limiting or reinforcing the loudness of vibrations has a large influence on the listener. Research into vibrations has shown that all vocalists produce a distinct vibration in the low and mid-range, but that the vibration characteristics are somehow degraded in the high-frequency range [1–3]. However, there are a few female sopranos who produce the highest notes and use vibrations even in the high frequency band. Therefore, in this paper, in order to confirm whether sopranos actually shows distinct vibrations even in the high—frequency range, the vibration characteristics of four Korean sopranos are analyzed with respect to their vibration depth. Chapter "Design and Evaluation of a MMO Game Server" of this study analyzes the production and characteristics of voices, Chapter "Automatic Generation of Image Identifiers Based on Luminance and Parallel Processing" explains the vocal characteristics of sopranos Chapter "Interface Module for Emulator-Based Web Application Execution Engine" discusses the experiment and it's results. Chapter "A Study on the Influence and Marketing Effect of KoreanWave Events and Festivals Organization" is the conclusion of this study.

2 Creation and Characteristics of Voices

2.1 Voice

Voice is a convenient and effective medium for humans to communicate, and consists of a sound wave in which air emitted from the lungs is radiated from the stenosis point so that the air is shaken or vibrated. Sound generated by this excitation source from the lungs emerges as speech [4–6]. The pitch (sound height) is determined according to the vibrations of the vocal cords and the amplified frequencies are different according to different body structures. The frequency amplified according to these structures is called the system characteristic, and the system characteristics influenced by the human body structure is called a formant. Formants are classified according to their frequency. The formant with the lowest frequency is known by convention as the first formant, followed in order of ascending frequency by the second, third, fourth and fifth formants. The following is a voice generation model that models the process of voice generation and is the basis of all voice related fields (Fig. 1).

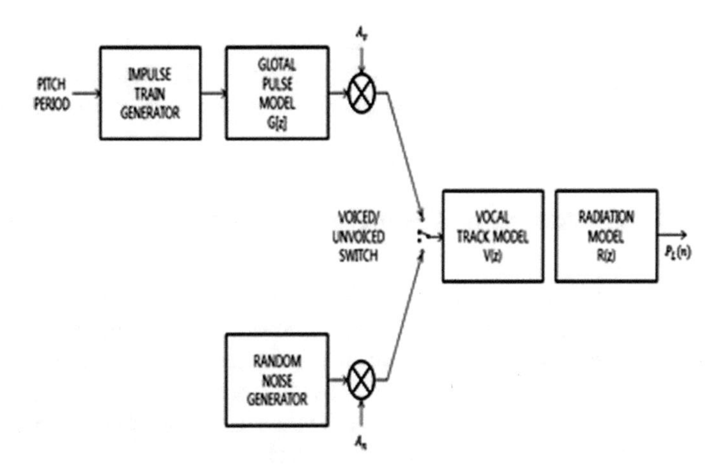

Fig. 1 Voice generation model

2.2 *Pitch*

Voiced sounds occur when air is forcibly thrust through the open part between the vocal folds. The tension of the vocal cords is adjusted to vibrate. The time from the opening of the vocal cord muscle to the opening of the next vocal cord muscle is called the fundamental period T0 and the speed at which the vocal cords tremble is called the fundamental frequency F0 = 1/T0. The basic cycle depends on the size and tension of the vocal fold of the person speaking at a given point in time. For example, on average, an adult male's vocal cord muscle is larger than an adult female's vocal cord muscle, so at the same frequency an adult male's voice is lower than that of an adult female. The term pitch is often used in the same sense as the fundamental frequency, but there is a subtle difference between the two. Everyone has a pitch range that is constrained by his larynx structure. For men, the possible pitch range is usually between 50 and 250 Hz, and between women between 120 and 500 Hz [7–9].

2.3 *Formant*

The formant will change its resonant frequency by movements of the cavities in the tongue and neck, depending on which voice is to be pronounced. Conversely, if we know the resonance frequency of the speech waveform from the lips, we can figure out which person pronounced the voice. Further, in the spectrum, resonant frequencies represent the peaks of the spectrogram. Generally speaking, the phonological properties of the phoneme are determined by the first formant (F1) and the second formant (F2), and F3, F4 and F5 contain the personality of the person. F1 or F2 are most important in vowels or tones, but in the case of fricatives, the formants are not always as obvious as outlined above, but they are very complex and a good amount of phonological information is found in F1, F2 as well as in F3, F4 and F5 [10–13].

3 Vocal Characteristics of Representative Soprano Singers

Female vocalists are classified into three types according to their vocal range: sopranos, mezzo-sopranos, and contraltos. The soprano has the highest vocal range. Because vocalization occurs through vocal organs according to the human voice production model, it is strongly influenced by the vocal organs. In particular, pronunciation characteristics are most affected by the size of the mouth, and when the mouth is small, the pronunciation becomes unclear.

3.1 Vocalization of Vocal Music

Vocalization occurs by vibrating the vocal cords during the breathing process. The respiration process can be roughly divided into two processes: inspiration and exhalation. Inhalation refers to the process in which the diaphragm descends and the volume of the lungs expands and external air enters the lungs due to the difference in density with the outside atmosphere. The breath is the volume of the lungs as the diaphragm returns to its original position, which means that the air of the lungs is released to the outside atmosphere by the recoil action. As such, good vocalization is highly dependent on good breathing skills, and as there are important differences between general breathing methods and the breathing methods for vocalization, vocalists constantly perform breathing and vocal training [14–17].

3.2 Voice Characteristic of Korean Representative Soprano

Female vocalists are divided into sopranos, mezzo-sopranos and altos according to the height of the transliteration. Sopranos are the vocalists who sing with the highest-pitched sounds. However as humans age, out voices tend to become lower. This is a fatal phenomenon for sopranos, who tend to have a shorter singing life span than other types of vocalists. However, and interestingly, an analysis of the singing voice of the internationally well-known soprano Jo Su-mi, who has been singing for over 30 years, shows that her voice has remained essentially the same for the past 20 years. Her voice also shows uniform vibration in the low, mid, and high frequencies, yet produces a distinctive sound. Further, Jo su-mi is also famous for lacking an uvula, which is not disturbed by the vibrations of the vocal cords, and makes her voice suitable for the role of the treble. Sopranos can be heard performing Coloratura, which has a treble above 1000 Hz, for example in such representative coloratura soprano roles as “Ariadne of Naxos Island”.

Figure 2 below is a partial spectrogram showing a certain interval of vibration in the ‘The Queen of the Night’ sung by Jo Su-mi.The spectrograph shows a constant vibration interval at 700, 800, and 1000 Hz, and it can be seen that even if the pitch of the sound changes, the vibration continues. Figure 3 below is a spectrogram of a part of the highest coloratura area of the ‘Ariadne of Naxos Island’, also sung by Cho Su Mi. The marked part shows a height of about 1400 Hz, which is a definite coloratura, higher than 1000 Hz. In addition, we can see in Fig. 3 the constant vibration according to the height of the note that we confirmed in Fig. 2, and it can be seen that the sound of 1400 Hz descends naturally low even after voicing for about four seconds.

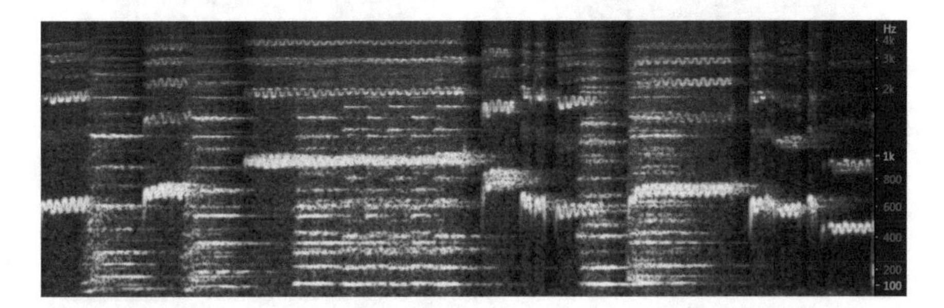

Fig. 2 Spectrogram of ‘The Queen of the Night’

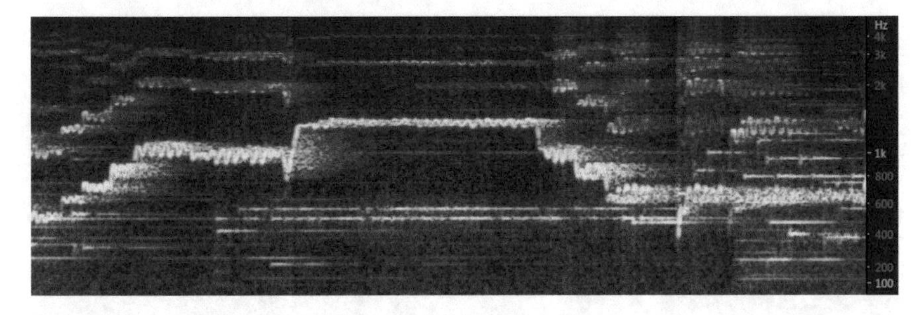

Fig. 3 Spectrogram of ‘Ariadne of Naxos Island’

4 Experiments and Results

We used Adobe Audition CC and Praat programs for the analysis in this paper. Using the Adobe Audition CC program, we sampled a portion of four Korean sopranos A, B, C, and D at 8000 Hz, quantized it to 16 bits, and detected the fundamental frequency using the Praat program.

Figures 4, 5, 6 and 7 show the waveforms and spectrograms of the four sopranos A, B, C, and D.

The maximum fundamental frequency of soprano A was 800 Hz, and the minimum fundamental frequency was 660 Hz. The maximum fundamental frequency

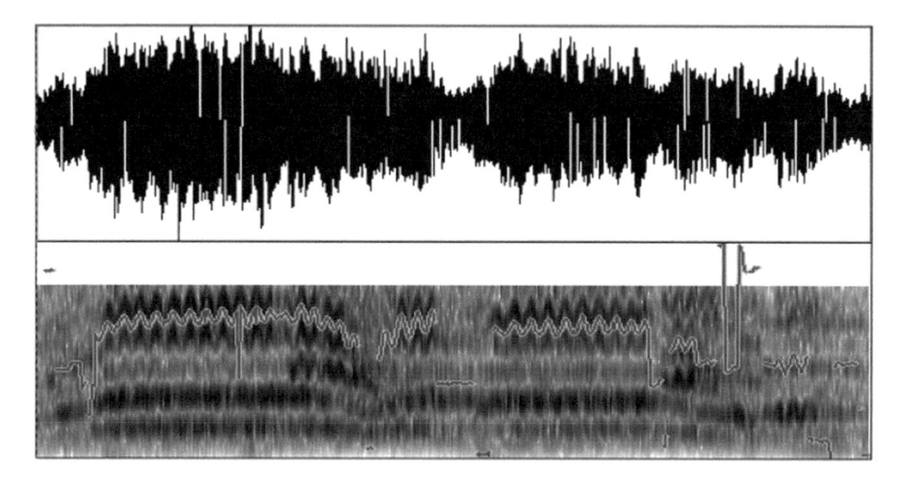

Fig. 4 Waveform, Spectrogram of Soprano A

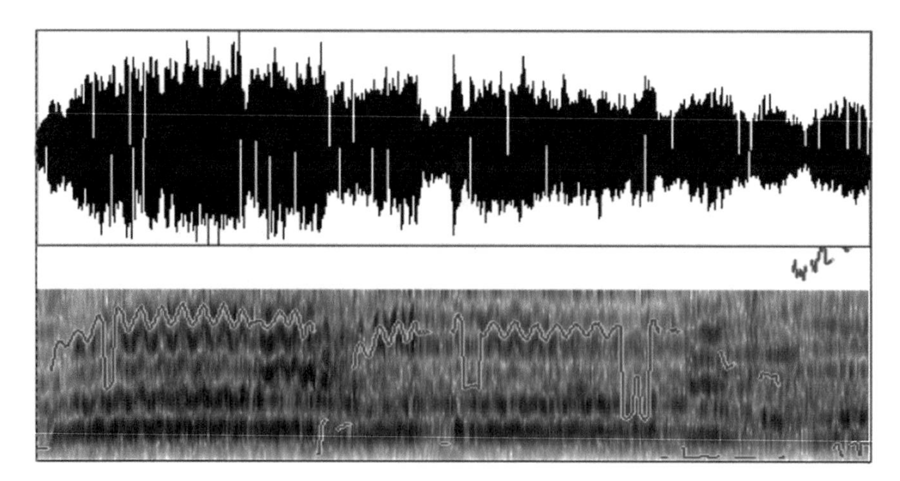

Fig. 5 Waveform, Spectrogram of Soprano B

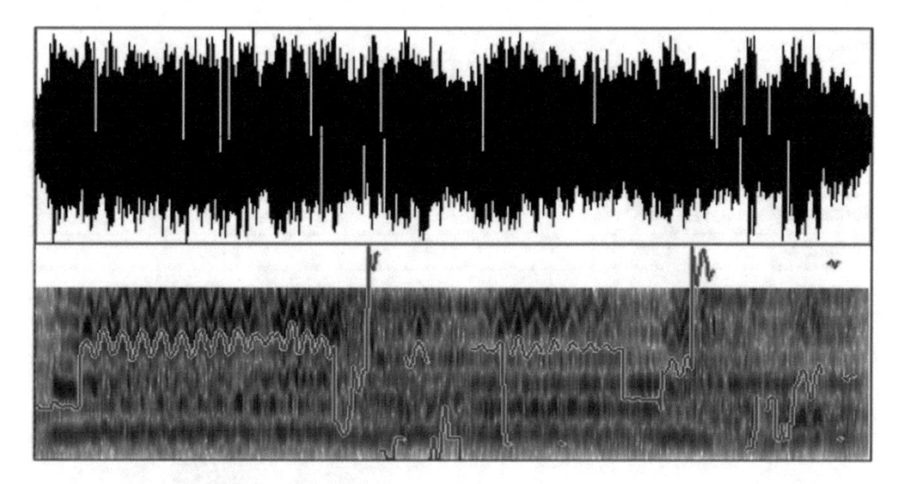

Fig. 6 Waveform, Spectrogram of Soprano C

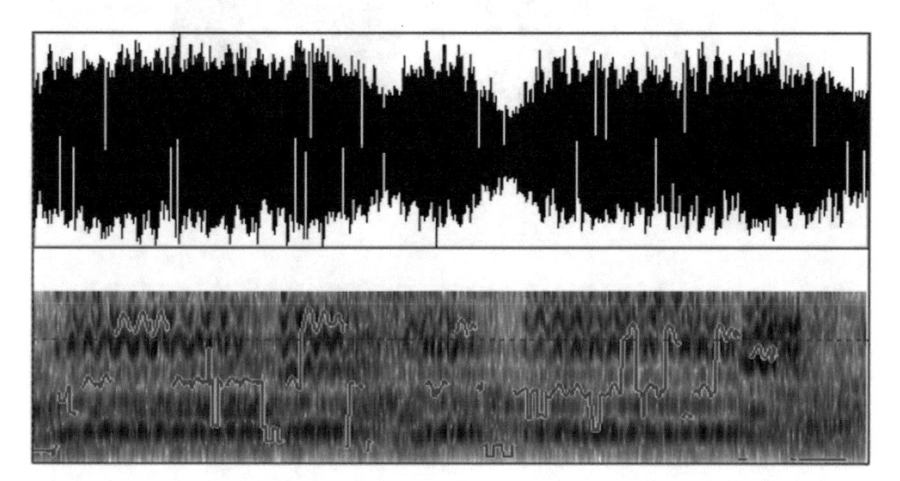

Fig. 7 Waveform, Spectrogram of Soprano D

of soprano B in the vibration range was 730 Hz, and the minimum fundamental frequency was 660 Hz. The maximum fundamental frequency of soprano C was 680 Hz and the minimum fundamental frequency was 540 Hz. Finally, the maximum fundamental frequency of the soprano D in the vibration range was 780 Hz, and the minimum fundamental frequency was 640 Hz. Table 1 above shows the measurement and analysis results (Fig. 8).

Finally, an MOS-TEST was conducted on ten healthy 20-year-old men to test the audibility of the four high-pitched and vibrating soprano singers A, B, C, and D. The test tells ten listeners in order that the soprano singer A 'If leave', the soprano singer B 'If leave', the soprano singer C 'If leave', the soprano singer D 'If leave', I tested my hearing. As a result of the test, the score of soprano A was 1.9, the score

Table 1 Measure by each Soprano Singer

	Maximum frequency (Hz)	Minimum frequency (Hz)	Variation
Soprano A	800	660	140
Soprano B	730	660	70
Soprano C	680	550	130
Soprano D	780	640	140

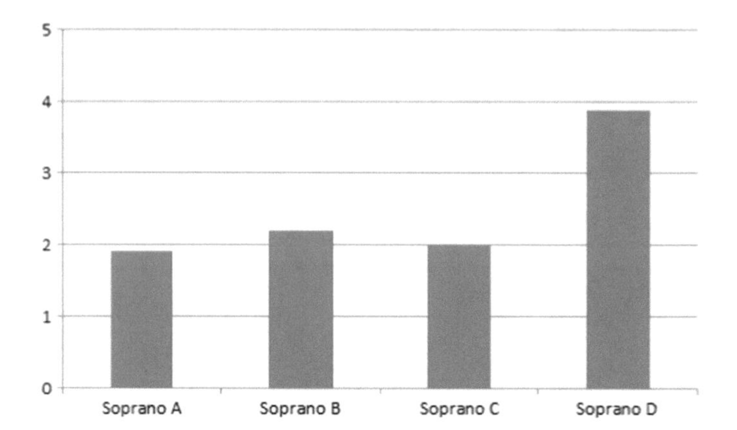

Fig. 8 MOS-TEST result

of soprano B was 2.2, the score of soprano C was 2.0, and the score of soprano D was 3.9. According to the results of the MOS-TEST, soprano D has the highest level of audibility, followed by soprano B, soprano C and soprano A in that order.

5 Conclusion

There are many different types of singers, such as popular music singers, Korean traditional musicians, and classical musicians. One of the most important elements of singing is vibration. In this paper, we analyzed the vibration characteristics of four Korean sopranos in order to confirm whether they show distinct vibrations even in the high frequency range. The vibration depth is a percentage of the change in the maximum fundamental frequency and the minimum fundamental frequency of each soprano in a certain vibration interval and the ratio of the soprano's total band as a percentage. The results are as follows: the maximum fundamental frequency is 800 Hz and the minimum fundamental frequency is 660 Hz in the vibrational section of soprano A; the maximum fundamental frequency of soprano B in the vibration range was 730 Hz, and the minimum fundamental frequency was 660 Hz; The maximum fundamental frequency of soprano C is 680 Hz, and the minimum fundamental frequency is 550 Hz; and the maximum fundamental frequency of soprano D in the vibration range was 780 Hz, and the minimum fundamental frequency was 640 Hz. As such, soprano singer D has the highest vibration depth, followed by soprano A, soprano C, and soprano B in turn, and the overall vibration depth is high.

References

1. Baek, G.-R., Yun, J.-J., Bae, M.-J.: A study on a similarity of a voice in the same lineage. Proc. Acoust. Soc. Korea **29**(2), 447–448 (2010)
2. Yi, E.-Y., Bae, M.-J.: A study on a color expression of sound. In: Conference Proceedings of Convergence Research Letter, Convergent with Art, Humanities, and Sociology, vol. 2, no. 4, pp. 1671–1674 (2016)
3. Kyon, D.-H., Bae, M.-J.: A voice similarity study about vocal organ proportion. In: Conference Proceedings of Voice Communication and Signal Processing, The Acoustical Society of Korea, vol. 28, no. 1, pp. 103–104 (2011)
4. Lee, W.H., Baek, G.R., Bae, M.J. (eds.): Valid-frame distance deviation of drunk and non-drunk speech. Korean Instit. Commun. Info. Sci. **53**, 876–877 (ISSN 2287-2639) (2014)
5. Kim, B.Y., Yi, E.Y., Bae, M.J.: A study on sound a color for various voices. In: Conference Proceedings of Convergence Research Letter, Convergent with Art, Humanities, and Sociology, vol. 3, no. 2, pp. 1135–1138 (2017)
6. Seo, D.-W., Sang B.-P., Bae M.-J.: A study on the characteristics of presidential voices similarity in four countries through voice analysis. In: Convergence Research Letter, vol. 3, no. 2, pp. 1155–1159 (ISSN: 2384-0870) (2017)
7. Park, S.-B., Lee, W.-H., Kim, M.-S., Bae, M.-J.: A study on classification of parameters for restoring speech signal. In: Convergence Research Letter, vol. 3, no. 4, pp. 1473–1476 (ISSN: 2384-0870) (2017)
8. Song, U.-J., Bae, M.-J.: A study on comparison of pronunciation characteristics of soprano vocalists. In: Advanced and Applied Convergence Letters, vol. 9, pp. 87–90 (2017)
9. Lee, W.-H., Bae, S.-G., Bae, M.-J.: A study on the possibility of drinking through speech waveform compensation in wireless communication environments. In: Advanced and Applied Convergence Letters, vol. 9, pp. 104–106 (2017)
10. Lee, Y.M.: ‘C’est si bon’ singing trot with Seotaiji. Durimedia, Korea (2011)
11. Choi, G.S.: Column of the ‘Nam in-su’s song, Column of weekly hankook, (2005)
12. Yoo, J.Y.: A Study of acoustic analyses of singers’ voice. Doctorate thesis, Graduate School, Taegu University, Korea (2003)
13. Park, H.W., Jee, S.H., Bae, M.J.: Study on the confidence-parameter estimation through speech signal. J. Multimedia Serv. Convergent Art, Humanit., Sociol. **6**(7), 101–108 (2016)
14. Yi, E.-Y., Kim, B.-Y., Bae, M.-J.: A study on sound characteristics of classical vocal musician using sound color, In: Convergence Research Letter, vol. 3, no. 3, pp. 1549–1552 (ISSN: 2384-0870) (2017)
15. Bae, J.H., Ahn, I.-S., Bae, M.-J.: On sound characteristics of bo gum park’s voice. In: Convergence Research Letter, vol. 3, no. 1, pp. 1037–1040 (ISSN: 2384-0870) (2017)
16. Lee, Y.M.: Trends and Issues in Trot Controversies, Criticism and Theory, The Criticism and Theory Society of Korea, vol. 13, no. 1, pp. 33–68 (2008)
17. Jung, C.J., Bae, M.J.: A study on a vocal characteristic of Lee mi-ja’s song. In: Proceedings of Symposium of The Acoustical Society of Korea, vol. 28, no. 1, pp. 41–42 (2009)

Author Index

R. Lee (ed.), *Software Engineering Research, Management and Applications*, Studies in Computational Intelligence 789, https://doi.org/10.1007/978-3-319-98881-8

Printed in the United States
By Bookmasters